FRACTAL GEOMETRY:
Mathematical Methods, Algorithms, Applications

Based on the proceedings of the First IMA Conference on 'Fractal Geometry: Mathematical Methods Algorithms and Applications'. Organised by the Institute of Mathematics and its Applications and held at De Montfort University, 20-22 September 2000

Co-sponsored by the Institute of Physics and the Institute of Electrical Engineers

The Institute of Mathematics and its Applications,
Catherine Richards House,
16 Nelson Street,
Southend-on-Sea,
Essex SS1 1EF
{http://www.ima.org.uk/}

The Institute of Physics,
76 Portland Place,
London
W1N 3DH
{http://www.iop.org/}

The Institution of Electrical Engineers,
Savoy Place,
London
WC2R OBL
{http://www.iee.org/}

Horwood Publishing Series: Mathematics and Applications

Cultural Diversity in Mathematics (Education)
Eds: A. Ahmed *et al*, University College, Chichester
Geometry with Trigonometry
P.D. Barry, National University of Ireland, Cork
Fractal Geometry: Mathematical Methods, Algorithms, Applications
J.M. Blackledge, A.K. Evans & M.J.Turner, De Montfort University, Leicester
Advanced Imaging: Theory, Computing, Practice
J.M. Blackledge & M.J.Rycroft, De Montfort University, Leicester
Digital Signal Processing: Software Solutions and Applications
J.M. Blackledge & M.J. Turner, De Montfort University
Image Processing III: Mathematical Methods, Algorithms & Applications
J.M. Blackledge *et al*, De Montfort University
Image Processing II: Mathematical Methods, Algorithms & Applications
J.M. Blackledge & M.J. Turner, De Montfort University
Electrical Engineering Mathematics
R. Clarke, Imperial College, London
Mathematical Modelling (ICTMA 8)
Eds: P. Galbraith *et al*, University of Queensland, Brisbane, Australia
Experimental Design Techniques in Statistical Practice
W.P. Gardiner & Gettinby, Glasgow Caledonian University
Delta Functions: Fundamental Introduction to Generalised Functions
R.F. Hoskins, De Montfort University
Teaching & Learning Mathematical Modelling (ICTMA 7)
Eds: S.K. Houston *et al*, University of Ulster, Northern Ireland
Calculus: Introduction to Theory and Applications in Physical and Life Sciences
R.M. Johnson, University of Paisley
Linear Differential and Difference Equations
R.M. Johnson, University of Paisley
Game Theory: Mathematical Models of Conflict
A.J. Jones, University College of Wales, Cardiff
Stochastic Differential Equations and Applications
X. Mao, University of Strathclyde
Teaching Practice in Mathematics
J.H. Mason, The Open University, Milton Keynes
Modelling and Mathematics Education (ICTMA 9)
J.F. Matos *et al*, University of Lisbon, Portugal
Manifold Theory
D. Martin, University of Glasgow
Fundamentals of University Mathematics, 2nd edition
C. McGregor *et al*, University of Glasgow
Mathematics for Earth Sciences
P. Sharkey, University of Portsmouth
Decision and Discrete Mathematics
The Spode Group, Truro School, Cornwall
Mathematical Analysis and Proof
D.S.G. Stirling, University of Reading
Statistical Mechanics: An Introduction
D.H. Trevena, University of Aberystwyth
Ordinary Differential Equations and Applications
W.S. Weiglhofer & K.A. Lindsay, University of Glasgow
Geometry of Navigation
R. Williams, Master Mariner

Fractal Geometry:
Mathematical Methods, Algorithms, Applications

Editors:
J.M. Blackledge, Institution of Simulation Sciences
A.K. Evans, Modern Optics Centre
M.J. Turner, Institution of Simulation Sciences
DeMontfort University, Leicester

Horwood Publishing *for* **The Institute of Mathematics**
Chichester **and its Applications**

Published in 2002 by
HORWOOD PUBLISHING LIMITED
Coll House, Westergate, Chichester, West Sussex, PO20 6QL England

British Library Cataloguing in Publication Data
A catalogue record of this book is available from the British Library

ISBN 1-904275-00-1

Contents

Index of Authors

Preface

This volume contains the proceedings of the First IMA Conference on Fractal Geometry: Mathematical Methods, Algorithms and Applications, that was held at De Montfort University in September 2000. The emphasis was given to work that related the mathematical exposure given to a problem and the practicalities required to create and implement an algorithm.

Eleven papers were presented, and finally accepted and published here, by researchers from many areas of Europe, Russia and America giving a multi-national emphasis to the fractal geometry problems described and the conference as a whole. Prof. Per Bak from the Neils Bohr Institute for Astronomy, Physics and Geophysics, while on sabbatical with the Department of Mathematics and Theoretical Physics at Imperial College, London, gave an entertaining keynote talk on being able to observe the global structure from local fractal geometric properties, under the title *Forest Fires, Measles, and the Structure of the Universe.*

Due to unforeseen circumstances a full transcript is unavailable, but a copy of the abstract is presented in full, "Forest fire models can be used to describe turbulent phenomena, where energy is injected at a large length scale and dissipated at a small scale. The fractal dimension of fires varies gradually from zero to three as the length scale is increased. A study of galaxy catalogues indicates that the distribution of luminous matter in the universe follows a similar pattern. At small distances, the universe is zero-dimensional and point-like; at distances of the order of 1 Mpc the dimension is unity, indicating a filamentary, string-like structure. When viewed at larger scales it gradually becomes two dimensional and pancake-like; finally, at the correlation length, 300 Mpc, it becomes uniform. The dissipative field in turbulence may follow a similar pattern. Moreover, the distribution of real forest fires in the US and Australia, and the distribution of measles epidemics mimics the distribution of fires in a modified version of the model due to Drossel and Schwabl." Further details are available in an extensive publication list of Per Bak's from the famous popular science book *How Nature Works* (5 June, 1997) Oxford University Press; ISBN: 0198501641, and *Self-similarity of extinction statistics in the Fossil Record* Ricard V. Solé, Susanna C. Manrubia, Michael Benton, and Per Bak. *Nature* 388, p.764, 21 August 1997 (Featured in *New Scientist*, 8 November 1997, and in *New York Times*, 2 September 1997), to the more directly relevant and recent works including *Scale Independent Dimension of Luminous Matter in the Universe* Per Bak and Ken Chen, *Phys. Rev. Lett.* 86 4215 (2001) and *Solitons in the one-dimensional forest fire-model* Per Bak and Ken Chen, *Phys. Rev. Lett.* 86 2475 (2001).

The proceedings presented here contain eleven of the presentations as full published papers. Some of the papers were not published, due to patent issues and it is hoped and planned to publish them in a following conference and proceedings when the IPR issues are fully resolved. The IMA is already planning to host the next conference on fractal geometry in September 2003 (details can found on the IMA web site `http://www.ima.org.uk/`). As is healthy, in any field, there is a good mixture of theoretical and applied papers.

The variation of applications of fractal geometry is immense, as the study has been embraced by many different scientific fields, and this is reflected here in their use from finance to geology and from aeromechanical design to encryption. To quote John Archibald Wheeler, protégé of Neils Bohr "No one will be considered scientifically literate tomorrow who is not familiar with fractals".

The editors would like to thank all those who contributed to the proceedings and those involved in organising the conference. We wish to show our gratitude to the continual support of the main sponsors; the Institute of Mathematics and its Applications and the co-sponsors; the Institute of Physics and the Institute of Electrical Engineers.

Jonathan M. Blackledge, Allan K. Evans, Martin J. Turner

Front Cover Image

The Institute of Simulation Sciences (ISS) currently consists of twelve research professors, five researchers and over 25 PhD students looking at a range of diverse areas from fractal coding techniques and noise modelling to shock wave implosion simulations and laser scattering.

The front cover shows a stylised version in the blue spectrum of an electrographic photograph of a leaf. This represents one of many projects being carried out by members of the Institute. The image was captured by discharging a 50KV Tesla coil charge and photographing the result using a Canon EOS10 with a non-coated lens captured on ISO 800 film, with an exposure time of about 5 seconds.

Electrography was developed mainly by the Armenian Semyon Davidovich Kirlian from 1939, although it had been known about before: Georg Lichtenberg in 1777 recorded electrical discharges in resin and after photography was developed, Batholomew Navratil created many examples and first used the term *electrography* in 1888. It is often known by the name Kirlian photography, and the corona has unfortunately been associated with unfounded 'parapsychological' causes.

The work is preliminary and future development is planned to apply fractal analysis to the corona within a microscope setup so that cell analysis can take place, and potential discriminations detected. It has been postulated that the corona may enhance the visual effects of cell boundary roughness.

IMA Short Course

On the preceding day to the conference, there was a short course on fractal geometry. The schedule of the presentations is shown below:

Basic Concepts: Geometry, Dimension and Self-Similarity
Prof J.M. Blackledge
Deterministic Fractals and Pathological Monsters *Dr M. Crane*
Stochastic Fractals and Noise *Dr M.J. Turner*
Chaotic Dynamical Systems and Fractals *Dr A. Evans*
Fractional Calculus and Fractals *Prof R. Hoskins*
Fractional Dynamics and Lévy Statistics *Prof J.M. Blackledge*
Nonlinear Systems Dynamics and Chaos *Prof M. Goman*
Fractals in Digital Signal and Image Processing *Dr M.J. Turner*
Fractals and Simulation *Prof J.M. Blackledge*

Institute of Simulation Sciences
SERCentre, Hawthorn Building,
De Montfort University,
Leicester
LE1 9BH UK

{http://www.serc.dmu.ac.uk/ISS}

www.dmu.ac.uk
DE MONTFORT UNIVERSITY
LEICESTER • BEDFORD

This document was typeset with the LaTeX document preparation system on a PC running the Linux operating system. There are approximately 65,000 words and over 320 PostScript images and diagrams. Out of the eleven papers some were converted from other formats and it is hoped that errors have not been introduced, but this cannot be guaranteed and the editors apologise in advance if this has occurred.

Jonathan M. Blackledge (Editor)

Jonathan Michael Blackledge is a graduate of Imperial College and completed his PhD in theoretical physics at London University in 1983. He continued at the University of London as Research Fellow of Physics from 1983 to 1988 specialising in inverse problems in electromagnetics and acoustics. In 1988, he joined the Department of Applied Mathematics and Computing at Cranfield University where he promoted postgraduate teaching and research in Industrial and Applied Mathematics, including CAD/CAM/CAE and DSP.

In 1994, he was appointed to his current position at De Montfort University, where he holds a Chair in Applied Mathematics, is Director of the Institute of Simulation Sciences and Head of External Income Generation for the Faculty of Computing Sciences and Engineering.

Professor Blackledge has published over 100 research papers and technical reports for industry, two industrial software systems, three books and is a technical consultant to a number of industries in the European Union. He has lectured widely to audiences composed of mathematicians, computer scientists and engineers in areas ranging from programming and software engineering to artificial intelligence.

His current research interests include computer aided geometric design, fractal geometry in digital imaging and financial cryptography.

Professor Blackledge holds Fellowships with the Institute of Physics, the Institute of Mathematics and its Applications, the British Computer Society, the Institute of Electrical Engineers, the Institute of Mechanical Engineers and the Royal Statistical Society. His principal additional interests are in Music and European History.

Allan K. Evans (Editor)

Allan Evans graduated in Physics from the University of Cambridge. He spent a year completing a masters in Physics at the University of Minnesota, working on research on single-electron devices with Professor Leonid Glazman, and then returned to Cambridge to begin a PhD. His PhD work was on the statistical mechanics of chaotic dynamical systems, and also included some work on asymptotics in quantum theory.

In 1995 he moved to De Montfort University as a research fellow, working in the Department of Mathematical Sciences and the Imaging Research Centre. At De Montfort he has published research work in a number of fields. One focus of interest has been sonoluminescence, where he has studied the stability of spherically converging shocks inside bubbles, and also the stability of the bubbles themselves. He has also published work on the dynamics of traffic accidents and on the behaviour of thin surface films of buckminsterfullerene, as well as on financial time series analysis.

More recently, he has become interested in digital holography and other areas of applied optics, and now heads the Centre for Modern Optics at De Montfort University. The Centre carries out contract research and consultancy in holography and other areas of applied optics. His current work involves digital holographic microscopy, and also research on novel laser devices using holographically written microstructures.

Martin J. Turner (Editor)

Martin John Turner is currently a Senior Lecturer at De Montfort University, and is Laboratory Manager for the multi-disciplinary Institute of Simulation Science (ISS) and Laboratory Manager for the Digital Arts Laboratory (DAL). He gained his PhD in the Computer Laboratory, at Cambridge University, on image coding.

As Head of the Centre for Digital Signal and Image Processing, research interests include all forms of digital and analogue imaging and signal processing techniques, with fundamental work being carried out on coding schemes, visualisation and compression conscious operations. New data storage structures and on-the fly image manipulation operations have been implemented. Further development has led to novel image visualisation techniques and the design of an intermediate 'semi-compressed contour tree' to enable fast image manipulations, used in various projects within the ISS. Research in these fields has resulted in a short-term Fellowship with British Telecom, a published book *Fractal Geometry in Digital Imaging* [ISBN: 0127039709 Academic Press, 1998] as well as other published papers.

His consultancy has included freelance coding to assist local companies dealing with the 8501 controller for chemical plant pump monitoring and the creation of first digital spot-welder. The design of GUIs on both high- and low-level platforms has been his particular interest.

Teaching has involved supervising undergraduates and postgraduates in mathematics, computer science and design, as well as a demonstrator teacher for external courses, involving the Royal Air Force and British Gas. Dr Turner is course coordinator for the MSc in digital signal and image processing, as well as supervising and advising the PhD and MSc students within the ISS which has over 20 research students and has hosted over 250 general extra-curricular research meetings.

He is a member of the Institute of Electrical Engineers, the British Computer Society, the Institute of Mathematics and its Applications and is the University liaison with the SIRA Technology Centre.

Chaotic Dynamics in a Simple Aeromechanical System

M.G. Goman, A.N. Khrabrov[†] and A.V. Khramtsovsky

ISS, SERC, Hawthorn Building, De Montfort University, Leicester LE1 9BH

[†]Central Aerohydrodynamic Institute (TsAGI), Zhukovsky, Russia

Abstract

Dynamics of a free-to-roll delta wing installed at high incidence in a wind tunnel is outlined using the experimental and mathematical modelling results. A simple analytical model applied allows us to simulate the multiattractor and chaotic dynamics observed in wind tunnel tests and thus to validate the used method for nonlinear and unsteady aerodynamics loads representation.

Key words: Wing-rock, body/vortices interaction, unsteady aerodynamic model, multi-attractor dynamics, chaotic motion.

1 Introduction

The birth of multi-disciplinary theory of chaos is probably the most impressive event in science for the last several decades. Absorbing ideas and methods from pure mathematics, dynamical systems theory, symbolic dynamics, nonequilibrium thermodynamics, fractal geometry it has already helped to explain many problems that in the past were found intractable [1].

Our everyday lexicon widened with new terms such as *deterministic chaos, strange attractor, fractal dynamics*, etc. Some problems as the Lorenz attractor, the logistic map bifurcation diagram, the Mandelbrot sets, etc., solved in the beginning of the theory formation became its typical labels and one can see them on the covers of numerous texbooks.

The main feature of chaotic behavior in different systems is mainly connected with high sensitivity to initial conditions due to exponential divergence of all trajectories lying on the attracting structure, which is normally bounded in the phase space. Discovery of stable chaotic attractors in 1960s by Ed.Lorenz in low dimensional dissipative systems was rather unexpected. And it happened about 70 years after Henri Poincaré, based on his study of three-body-problem in celestial mechanics (conservative system), predicted, that: "*...it may happen that small differences in the initial conditions produce very great ones in the final phenomena. A small error in the former will produce an enormous error in the future. Prediction* *becomes impossible..." ([1]).* May be due to strong belief that in dissipative systems there exist stable attractors only in the form of equilibria, closed-orbits and tori this statement of Henri Poincaré had not been properly heard for such a long time. And only the appearance of computers has led to rediscovering this phenomenon. Modern computers enable us to apply, for chaotic dynamics investigations, qualitative methods [5], which provide geometrical and topological insight into nonlinear dynamics.

In this paper the problem of the 'wing rock' phenomenon arising as a limiting factor in high incidence flight of modern manoeuvrable aircraft is discussed in terms of chaotic behavior onset.

A simple aeromechanical system in the form of a free-to-roll delta wing mounted on a fixed sting in a wind tunnel (see Fig.1) at certain conditions can display chaotic dynamics, which is important in terms of verification of aerodynamics modelling method in the presence of vortical and separated flow conditions.

A comprehensive review of numerous papers devoted to investigation of the 'wing rock' phenomenon can be found in [2] with more than a hundred references. These works range from 'wing rock' observation in flight tests and experimental study in wind tunnel to empirical and computer fluid dynamics modelling. The 'wing rock' oscillations inhere practically in all modern fighters and can occur at high angles of attack at subsonic speeds and at flight with transonic speeds with relatively low angles of attack.

The identified causes of 'wing rock' motion are connected with the interaction of rigid body motion with vortical and separated flow processes, having their own internal dynamics. The latter is especially important, because the classical approach to aerodynamic loads modelling in flight mechanics is based on so called aerodynamic derivatives concept and practically ignores the internal flow dynamics.

Flight at high incidence of modern and advanced aircraft still demands an adequate method for mathematical modelling of nonlinear and unsteady aerodynamics necessary for dynamics simulation and control system design [3, 4]. The 'wing rock' modelling, even in a form of one-degree-of-freedom system, provides valuable opportunity for assessment of the aerodynamic model validity. A simple mathematical model for the 'wing rock' system displaying multi-attractor dynamics and chaotic behavior is analyzed in this paper and qualitatively compared with wind tunnel experimental data.

2 Experimental results

A series of delta wings with different sweep angle $65^o, 70^o, 80^o$ and double delta wing $80/60^o$ have been tested on a free-to-roll fixed sting in TsAGI's low speed wind tunnel. They demonstrated various types of motion such as stable equilibrium at zero bank angle, symmetrical stable large amplitude periodic oscillations, asymmetrical equilibria, asymmetrical periodic oscillations and also dif-

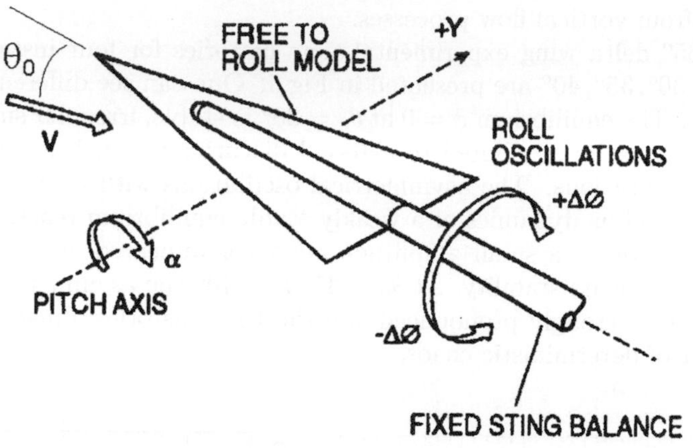

Figure 1. Schematic description of a free-to-roll experiment (after Katz [2]).

ferent forms of chaotic dynamics. Some wings at certain conditions displayed multi-attractor dynamics.

In Fig.2 free-to-roll 80° delta wing time histories for bank angle and phase portraits in the plane of the rolling moment coefficient versus bank angle are presented for two different initial conditions, i.e. $\phi(0) = 8^o$ and $\phi(0) = 24^o$, at $\theta_0 = 36^o$. In the first case the wing oscillates in a chaotic manner with small amplitude around zero bank angle. The rolling moment coefficient jumps between two branches of nonlinear dependency with a hysteresis between them. At larger $\phi(0)$ one can see the onset of the large amplitude limit cycle and linearization of the rolling moment coefficient dependency due to time lag effects. So, in this case the wing has two stable oscillatory attractors.

In Figs.3-5 experimental results for free-to-roll 65° delta wing are presented for different installation angles $\theta_0 = 20^o, 30^o, 35^o, 40^o$. If at $\theta_0 = 20^o$ the wing has a stable zero equilibrium (Fig.3, left), there are already three stable equilibria at $\theta_0 = 30^o$: $\phi = 0, \pm 22^o$ (Fig.3, right). At $\theta_0 = 35^o, 40^o$ the wing is converged to one of the two assymmetrical attractors in the form of irregular oscillations (Fig.4).

Analysis of the phase portrait projections of the experimental trajectories helps us to make an important conclusion about the dynamics model. For example, at $\theta_0 = 30^o$ (Fig.3, left) the phase trajectories, projected on the plane $(\frac{d\phi}{dt}, \phi)$, approach three different attractors and intersect each other. This means that normally used (see [2]) the second order dynamical system, describing just rigid body motion with aerodynamic moment based on classical aerodynamic derivatives representation, is not consistent. The intersection of the phase trajectories signifies that there is additional internal dynamics in this system and it comes from vortical flow processes.

The 65° delta wing experimental time histories for four installation angles $\theta_0 = 20^o, 30^o, 35^o, 40^o$ are presented in Fig.5. One can see different character of dynamics. The equilibrium $\phi = 0$ at $\theta_0 = 20^o$ is stable, irregular small amplitude oscillations can be attributed to external disturbances such as flow turbulence and sting vibrations. The asymmetrical oscillations with $\phi \approx 22^o$ at $\theta_0 = 30^o$ can be treated as dynamics of a weakly stable equilibrium reacting on external disturbances or as a small amplitude attractor appeared after an asymmetric equilibrium lost its stability. At $\theta_0 = 35^o$ and 40° the asymmetrical oscillations become more strongly pronounced and the time histories already resemble the behaviour of deterministic chaos.

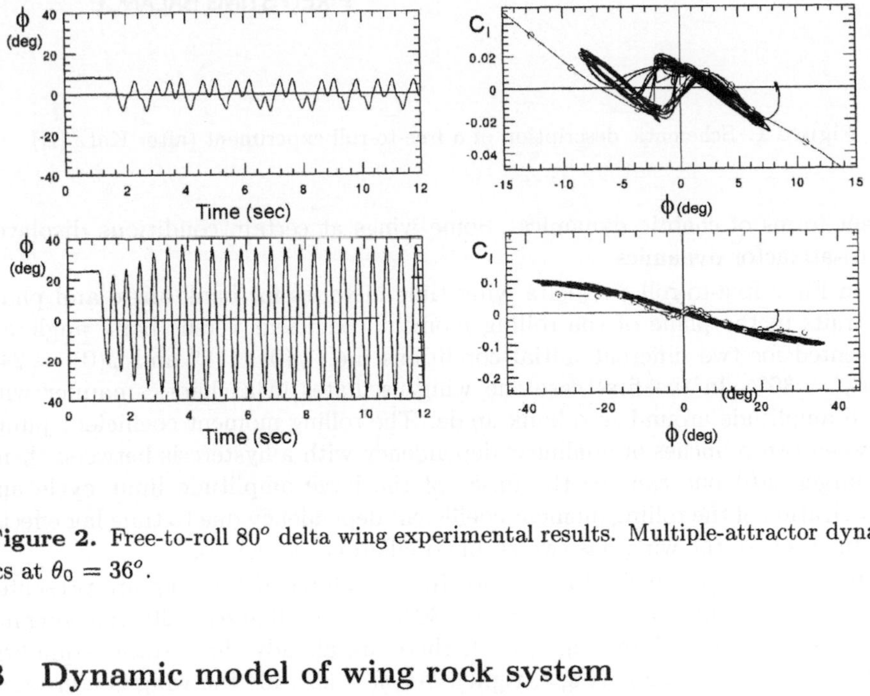

Figure 2. Free-to-roll 80° delta wing experimental results. Multiple-attractor dynamics at $\theta_0 = 36^o$.

3 Dynamic model of wing rock system

Kinematics of a free-to-roll delta wing installed on a fixed sting at pitch angle $\theta = \theta_0$ to the wind tunnel flow is expressed through the links between angle of attack α and sideslip β and bank angle ϕ:

$$\begin{aligned} \tan\alpha &= \tan\theta_0 \cos\phi \\ \sin\beta &= \sin\theta_0 \sin\phi \end{aligned} \tag{3.1}$$

The variation in bank angle $\phi(t)$ produces changes in angle of attack and sideslip $\alpha(t)$, $\beta(t)$, which specify the aerodynamic loads.

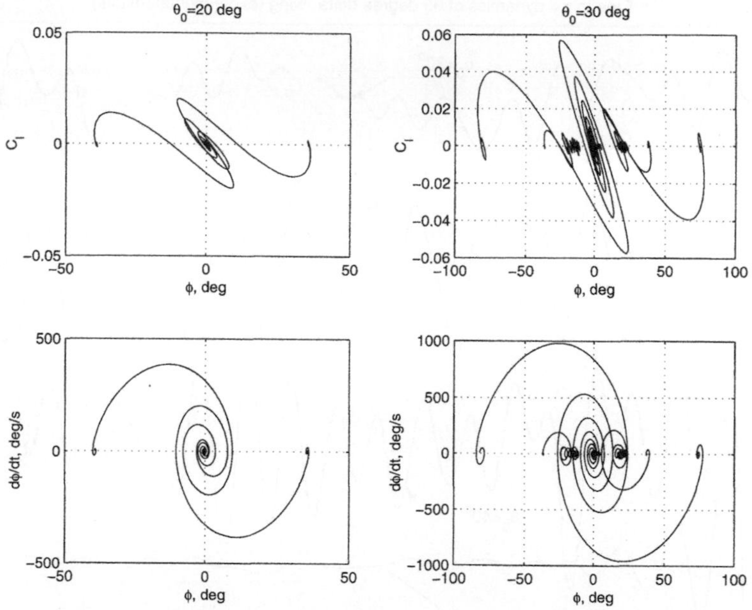

Figure 3. Free-to-roll 65° delta wing experimental results. Single attractor at $\theta_0 = 20^o$ (left) and multiple-attractor dynamics at $\theta_0 = 30^o$ (right).

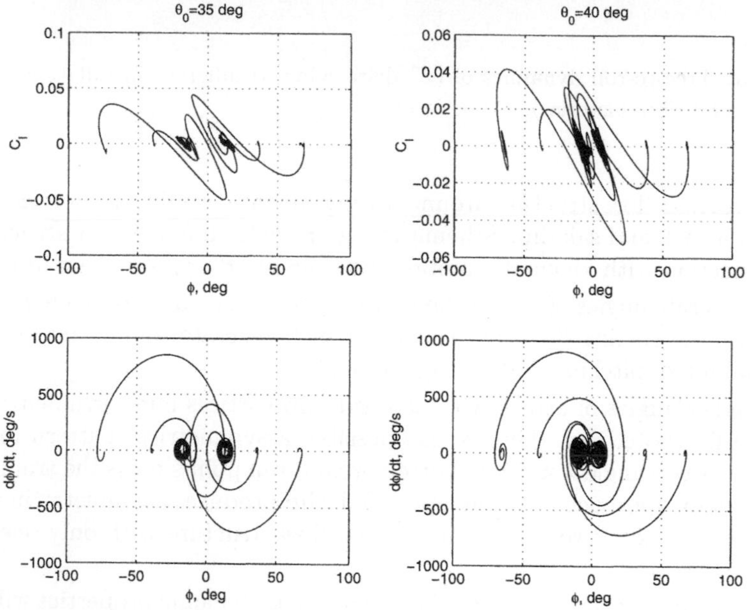

Figure 4. Free-to-roll 65° delta wing experimental results. Multi-attractor dynamics at $\theta_0 = 35^o$ and 40°.

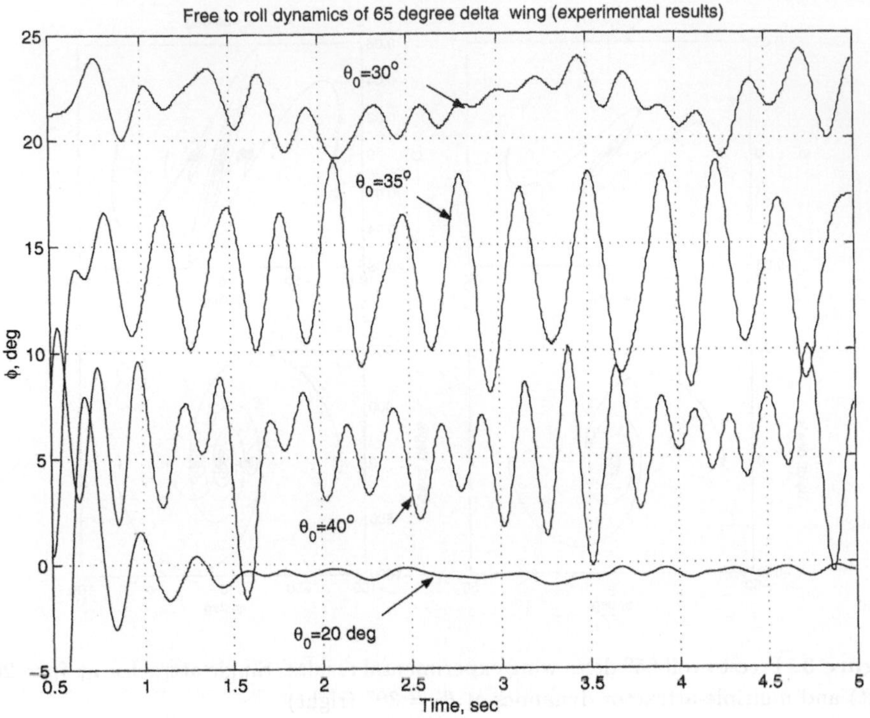

Figure 5. Free-to-roll dynamics of 65° delta wing at different installation angles θ_0 (experimental time histories).

The vortical flow structure around a wing in static conditions depends on the angles of attack and sideslip. Schematic regions with different flow structures [3] and connection with kinematic motion parameters (θ_0, ϕ) are given in Fig.6.

At moderate angles of attack the vortical sheets shedding from a wing leading edges form the vortical flow above a wing, which generates significant increments in normal force, pitching and rolling moments.

At higher angles of attack with zero sideslip vortices burst symmetrically at zero sideslip (note, that there is also possible assymmetrical pattern for wings with high sweep angle $\lambda \approx 80^0$). Vortex breakdown points cross the trailing edge and gradually approach the wing apex. Sideslip produces asymmetry in vortices and there also exist two regions of a mixed flow structure with only one, left or right, burst vortex.

During wing motion the vortical flow due to its dynamic properties will adjust to its shape and distribution at static conditions with some delay and thus produce unsteady aerodynamic loads, which can not be treated dependent only on

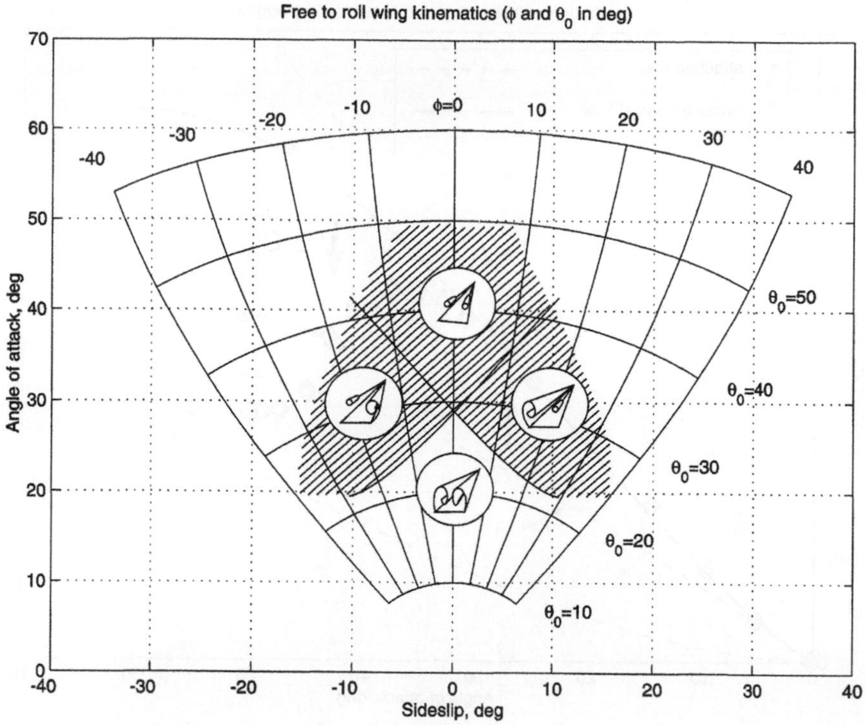

Figure 6. Free-to-roll wing kinematics and different flow structures.

wing states variables. To take this effect into account some additional dynamic states for internal flow dynamics have to be introduced.

The appropriate structure of the mathematical model for aerodynamic loads in such flow conditions has been proposed in [3, 4] based on a partitioning method. In Fig.7 the experimental dependency for the normal force coefficient on angle of attack is given along with the approximate dependencies computed using simple flow models such as attached flow ((1)-vortex lattice method), unburst vortical flow ((2)-vortex suction analogy of Polhamus) and fully stalled flow ((3)-conical approximation of Kirchgoff model with the region of constant pressure above the wing). The vortices generated on the leading edges of the delta wing produce significant increase in the normal force ΔC_{N_v} with respect to the attached flow case (1). The onset at $\alpha \approx 30^o$ and further development of vortex breakdown lead to the loss of the normal force $\Delta C_{N_{vb}}$ with respect to unburst vortical flow (2). Finally, after the vortex burst locations reach the apex of the wing ($\alpha \approx 45$ deg) the aerodynamic load approaches the curve defined by the fully stalled flow (3). Such simple analysis reveals the possible varia-

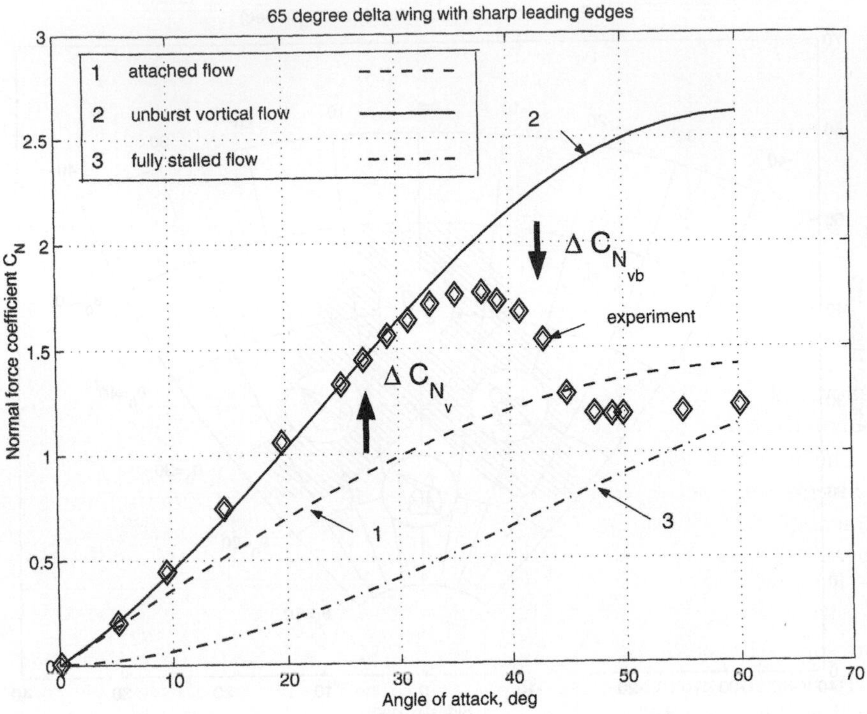

Figure 7. Aerodynamic loads at different flow structures. The normal force coefficient for the 65^0 delta wing: 1) attached flow, 2) unburst vortical flow, 3) fully separated flow, "diamonds" - experimental data.

tions in the aerodynamic loads due to flow structure changes. As long as the vortex breakdown and flow separation processes display big delays in flow adjustment the unsteady aerodynamic responses will be essentially non-linear and amplitude/frequency dependent.

The relatively slow dynamics of the vortex burst point produces considerable time lags in aerodynamic loads, which are much bigger than the convective time scale. The contribution to aerodynamic loads from vortex breakdown ΔC_{vb} therefore should be described by some dynamical model explicitly representing characteristic time scales defining the adjustment processes.

Similar decomposition of the static rolling moment coefficient on inertialess contribution from attached flow and dynamic contribution from vortical flow are presented in Fig.8, where a simple analytical representation of vortical flow effect is considered.

The rolling moment coefficient is expressed by a sum of attached flow contribution which depends only on the wing state variables (β - sideslip, p - roll rate) and vortical flow contribution C_{l_v}, which is dynamical:

$$C_l = C_{l_{\beta_0}}\beta + C_{l_{p_0}}\frac{pb}{2V} + C_{l_v},$$

$$\tau_1\frac{dC_{l_v}}{dt} + C_{l_v} = \frac{k_2 C_{l_{\beta_0}}\beta}{1+\beta^2/\beta_m^2} \tag{3.2}$$

The static dependencies for the rolling moment coefficient

$$C_{l_{stat}} = C_{l_{\beta_0}}\beta\left(1+\frac{k_2}{1+\beta^2/\beta_m^2}\right) \tag{3.3}$$

is shown in Fig.8 for $C_{l_{\beta_0}} = -0.4$ and different $k_2 = -5, -1, 0, 1$ ($\beta^* = \beta/\beta_m$ and conditionally $\beta_m = 1$).

The nonlinear dependency (3.3) at $k_2 > 0$ resembles the rolling moment coefficient at angles of attack with unburst vortical flow, at $k_2 < 0$ the rolling moment dependency is like in the case of vortex breakdown. At $k_2 < -1$ there appear two asymmetrical wing equilibria $\pm\beta_0$ extra to equilibrium $\beta = 0$ (see Fig.8). To take into account effect of the critical states crossing the nonlinear term in dynamic equation for vortical contribution C_{l_v} in (3.2) should contain more sophisticated nonlinearities, however, we will not consider this in the paper.

To close the mathematical model of a wing system with one degree-of-freedom in roll we have to add dynamic equation for rigid body motion:

$$\frac{d^2\phi}{dt^2} = \frac{\rho V^2 Sb}{2I_{xx}}C_l, \tag{3.4}$$

where C_l is the rolling moment coefficient, ρ is an air density, V is flow velocity, S and b are wing area and span, I_{xx} is wing moment of inertia.

Equations (3.1),(3.2) and (3.4) after converting them to dimensionless variables form the following system ($\phi \ll 1$):

$$\begin{aligned}
\frac{d\beta^*}{d\tau} &= p^*\sin\theta_0 \\
\frac{dp^*}{d\tau} &= k\left(C_{l_{\beta_0}}\beta^* + C_{l_{p_0}}p^* + C^*_{l_v}\right) \\
\tau_1\frac{dC^*_{l_v}}{d\tau} &= -C^*_{l_v} + \frac{k_2 C_{l_{\beta_0}}\beta^*}{1+\beta^{*2}},
\end{aligned} \tag{3.5}$$

where $\tau = \dfrac{tV}{b/2}$ is dimensionless time, $\beta^* = \dfrac{\beta}{\beta_m}$, $p^* = \dfrac{pb}{2V\beta_m}$, $C^*_{l_v} = \dfrac{C_{l_v}}{\beta_m}$ are dimensionless state variables, $k = \dfrac{\rho}{8\rho_w d\bar{r}_g^2}$ is dimensionless parameter, τ_1 is

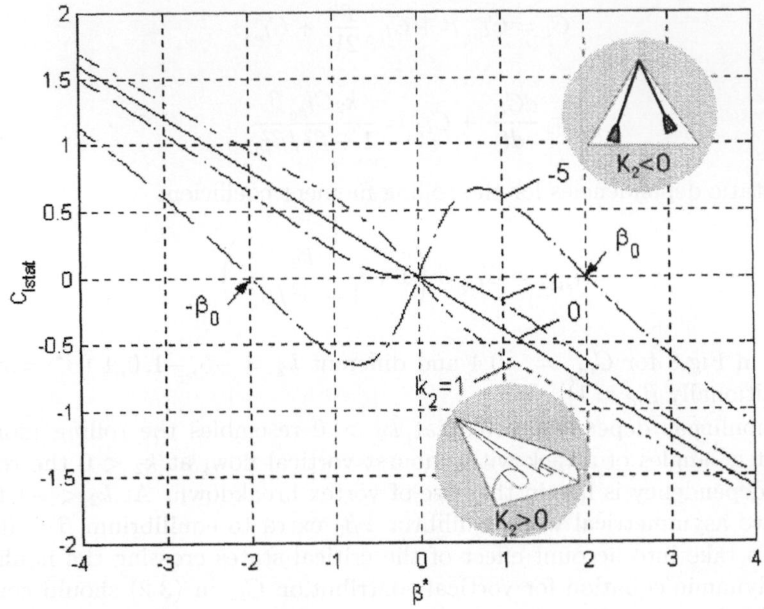

Figure 8. Rolling moment dependencies produced by vortical flow: unburst vortices - $k_2 > 0$, burst vortices - $k_2 < 0$.

characteristic time scale of vortical flow adjustment, ρ_w is a wing density, $\bar{d} = \dfrac{d}{b}$ is reduced wing thickness, $\bar{r}_g = \dfrac{\bar{r}_g}{b}$ is reduced wing radius of gyration.

Parameters k, k_2, τ_1, β_m vary with installation angle θ_0, wing characteristics such as sweep angle, shape of leading edges, mass distribution and density. For example, $k \in (0.002, 0.02)$, $k_2 \in (-10, 1)$, $\tau_1 \in (10, 30)$, $\beta_m \in (0.01, 0.25)$. So, by selection of these parameters from the above ranges one can analyze various types of wing dynamics.

4 Qualitative analysis of wing rock dynamics

Dynamical system (3.5) belongs to a class of oscillatory systems with inertial self-excitation [6]. The famous Lorenz equations belong to this class too.

In this paper we confine our analysis only by computation of bifurcation diagram against parameter k_2 using direct numerical simulation and mapping method. At every fixed value of parameter k_2 the numerical simulation is per-

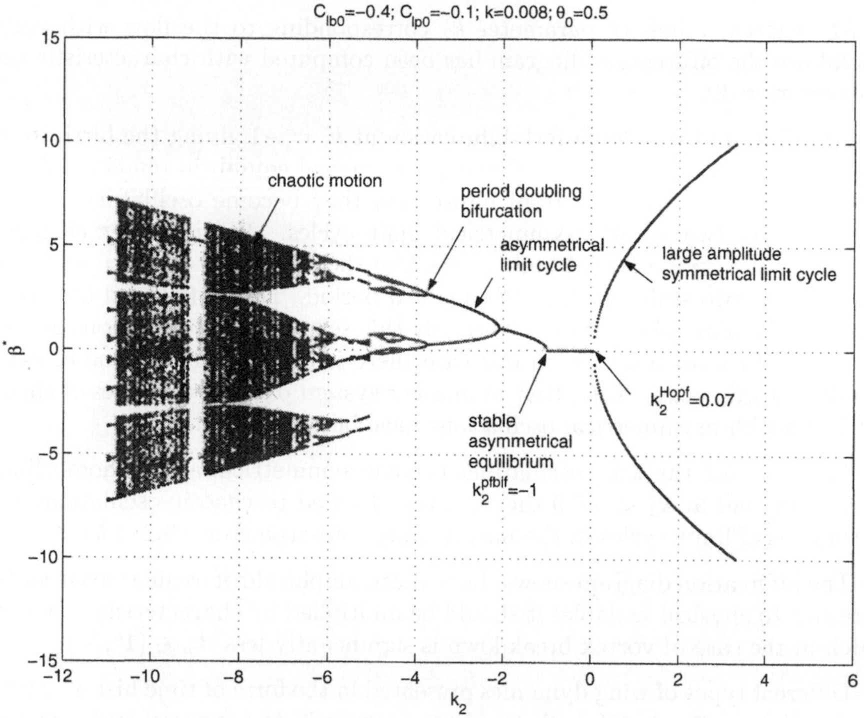

Figure 9. Bifurcation diagram of wing rock dynamical system (5).

formed on the time interval sufficient for approaching the steady dynamics and only mapping points on the plane $p^* = 0$ are plotted.

Gradual variation of the parameter k_2 allows us to compute a bifurcation diagram reflecting only stable attractors. In bifurcation diagrams computed in such a way a stable equilibrium is defined by only one point, a limit cycle is represented by two, four, eight and more separated points, and finally a chaotic attractor appears as a continuous distribution of points.

Fig.9 shows the bifurcation diagram versus parameter k_2 computed for system (3.5) with mapping condition $p^* = 0$ at $k = 0.008$ and $\theta_0 = 0.5$, $C_{l_{\beta_0}} = -0.4$, $C_{l_{p0}} = -0.1$.

At positive values of parameter k_2 the Hopf bifurcation takes place at $k_2 = 0.07$ giving the birth to symmetrical stable limit cycle. The reduced amplitude of a limit cycle after the Hopf bifurcation may be rather large $\beta^* \approx 5$ (note, that to compute the physical amplitude β it should be multiplied by $\beta_m \in (10^o, 20^o)$). Onset of such a limit cycle is typical for 80° delta wing. Note, that the computation in this case has been made with characteristic time constant

$\tau_1 = 10$.

At negative values of parameter k_2 corresponding to the flow with vortex breakdown the bifurcation diagram has been computed with characteristic time constant $\tau_1 = 32$.

The diagram has a pitch-fork bifurcation at $k_2 = -1$ giving the birth to two stable asymmetrical equilibria. These asymmetrical equilibria remain stable up to $k_2 = -2.1$ and after the Hopf bifurcation they become oscillatory unstable with onset of two stable asymmetrical limit cycles. With further change of parameter the asymmetrical limit cycles lose their stability at $k_2 = -3.75$ and there appear two stable cycles with doubled period. This is followed by a series of period-doubling bifurcations. Normally this sequence of bifurcations leads to the onset of chaos, however, in our case there is only a narrow zone of multi-doubled closed cycles. Note, that at another system parameters zones of chaotic dynamics with asymmetrical oscillations have been observed.

At $k_2 \leq -5.1$ the wing oscillations become symmetrical after a homoclinical bifurcation and at $k_2 < -5.9$ they are transformed to chaotic oscillations with restoration of limit cycles in the narrow range of parameter $-9.2 < k_2 < -8.72$.

The bifurcation diagram shows the reduced amplitude of oscillations in sideslip, to return to physical variables it should be multiplied by characteristic value β_m, which in the case of vortex breakdown is significantly less $\beta_m \in (1^o, 5^o)$.

Different types of wing dynamics presented in the form of time histories for all state variables β^*, p^*, $C^*_{l_v}$ and the phase portrait in the space of state variables are shown in Fig.10-12 for $k_2 = 0.5, -4.5, -7.0$ respectively.

The build-up of limit cycle oscillations at $k_2 = 0.5$ is seen in Fig.10. Such periodic symmetrical oscillations are typical for 80° delta wing with unburst leading edge vortices at installation angle $\theta_0 = 30^o$.

Two asymmetrical periodic attractors at $k_2 = -4.5$ after several period doubling bifurcations are shown in Fig.11, the oscillations in this case look like irregular ones.

And finally in Fig.12 one can see chaotic dynamics at $k_2 = -7.0$. Two unstable equilibrium points at positive and negative bank angle behave like a repeller and attractor simultaneously, so the trajectory jumps from one side to another in a nonregular chaotic manner.

In mathematical model (3.5) the rolling moment coefficient has been simplified in comparison with the real experimental dependency (minor nonlinearities have been omitted) and as a result the case with three different attractors (Fig.3, right) is lost in the model. However, the proposed mathematical model captures the main qualitative features of the wing rock motion observed in the wind tunnel experiments for different delta wings, i.e build-up of large amplitude symmetrical oscillations (see Fig.10 and Fig.2), irregular asymmetrical oscillations (see Fig.11 and Fig.5) and small amplitude chaotic oscillations (see Fig.12 and Fig.2).

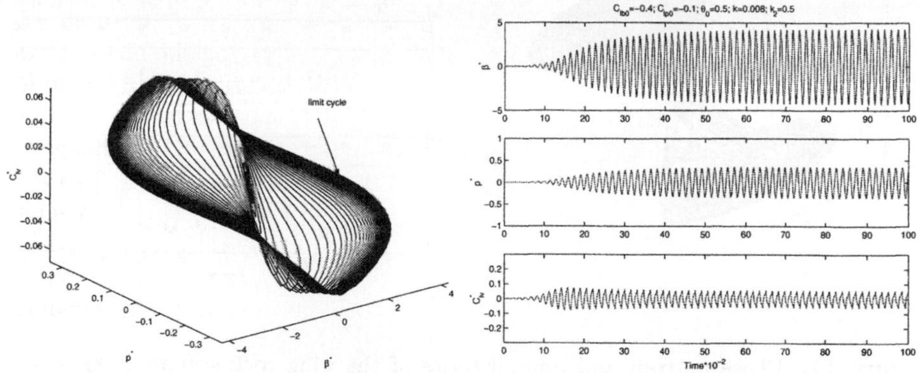

Figure 10. Phase portrait and time histories of the wing rock system at $k_2 = 0.5$. Build up of limit cycle oscillations.

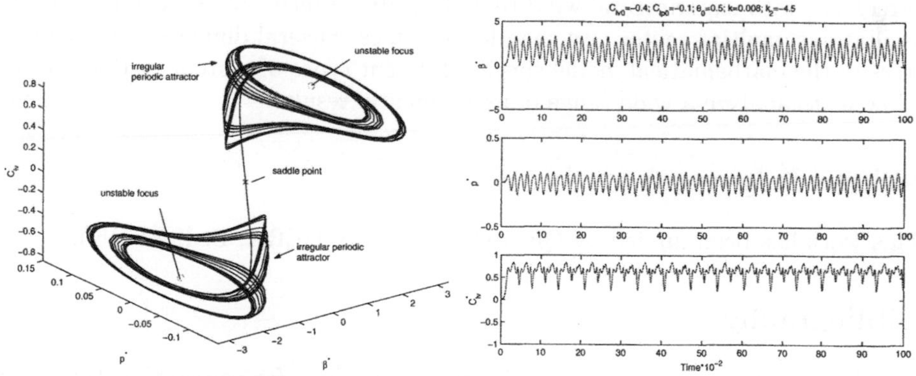

Figure 11. Phase portrait and time histories of the wing rock system at $k_2 = -4.5$. Irregular asymmetrical oscillations.

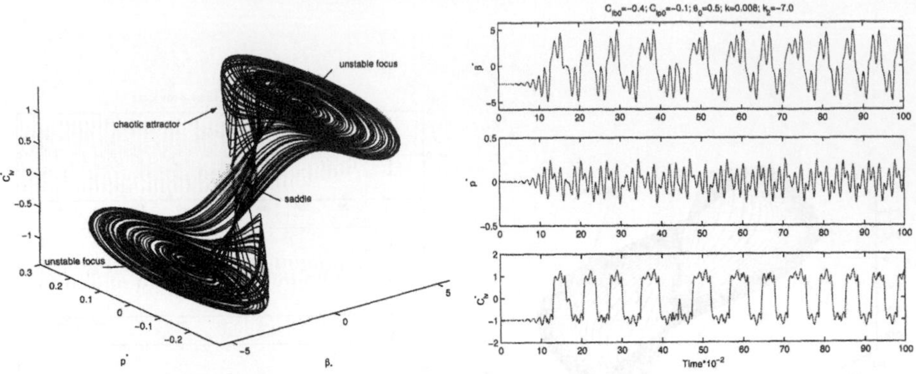

Figure 12. Phase portrait and time histories of the wing rock system at $k_2 = -7$. Chaotic symmetrical oscillations.

Concluding remarks

The experimental and mathematical modelling results for a free-to-roll delta wing demonstrate the importance of dynamic interaction between the wing and vortical flow processes, i.e. the importance of nonlinearities and time lag effects in aerodynamic loads. The proposed mathematical model for a free-to-roll delta wing installed on a fixed sting in a wind tunnel belongs to a class of oscillatory systems with inertial self-excitation. It allows us to obtain good qualitative agreement with experimental wind tunnel results in terms of onset of limit cycle oscillations, multiattractor and chaotic dynamics. Several dimensionless parameters in the mathematical model specify different types of system motion and so allow us to analyze a wide range of experimental results.

Acknowledgement

This work has been funded by QinetiQ (previously DERA, Bedford UK).

Bibliography

1. A.B. Cambel, *Applied Chaos Theory. A Paradigm for complexity*, Academic Press Inc., NY, 1993.

2. J. Katz, Wing/vortex interaction and wing rock, *Progress in Aerospace Sciences* 35 (1999), pp.727-750, Elsevier Science Ltd.

3. N. Abramov, M. Goman, A. Khrabrov and K. Kolinko, Simple Wings Unsteady Aerodynamics at High Angles of Attack: Experimental and Modeling

Results, Paper N 99-4013, *AIAA Atmospheric Flight Mechanics Conference*, August 1999, Portland, OR.

4. N. Abramov, M. Goman, D. Greenwell and A. Khrabrov, Two-Step Linear Regression Method for Identification of High Incidence Unsteady Aerodynamic Model, Paper N 2001-4080, *AIAA Atmospheric Flight Mechanics Conference*, August 2001, Montreal, Canada.

5. M. Goman, G. Zagainov and A. Khramtsovsky, Application of Bifurcation Methods to Nonlinear Flight Dynamics Problems, *Progress in Aerospace Sciences* 33 (1977), pp. 539-586, Elsevier Science Ltd.

6. Yu.I. Neimark and P.S. Landa, *Stohastic and Chaotic Oscillations*, Nauka, Moscow, 1987 (in russian).

Random walks with fluctuating step number, scale invariant behaviour and self-organised criticality

K.I. Hopcraft, E. Jakeman, R.M.J. Tanner

Theoretical Mechanics Division, School of Mathematical Sciences, University of Nottingham, Nottingham, NG7 2RD

Abstract

Novel diagnostics applied to a rice-pile cellular automaton reveal different mechanisms producing power-law behaviour of statistical attributes of grains which are germane to Self Organised Critical phenomena. The probability distributions for these quantities can be derived from two distinct random walk models that account for correlated/clustered behaviour through incorporating fluctuations in the number of steps in the walk. The first model describes the distribution for a spatial quantity, the resultant flight-length of grains. This has a power law tail caused by grains moving through a discrete, power-law distributed number of random steps of finite length. Developing this model into a random walk obtains distributions for the resultant flight-length having characteristics similar to Lévy distributions. The second random walk model is devised to explain a temporal quantity, the distribution of 'trapping' or 'residence' times of grains at single locations in the pile. Diagnostics reveal that the trapping time can be constructed as a sum of 'sub-trapping times', which are described by a Lévy distribution where the number of terms in the sum is a discrete random variable accurately described by a negative binomial distribution. The infinitely divisible, two-parameter, limit-distribution for the resultant of such a random walk is discussed and describes dual-scale power-law behaviour if the number fluctuations are strongly clustered. The form for the distribution of transit times of grains results as a corollary. The relationship between the dual-scale distribution and other stable and infinitely divisible distributions is elucidated.

1 Introduction

Interest in Self Organised Criticality (SOC) is multidisciplinary and continues to burgeon [e.g. 1]. The concept of SOC refers to the spontaneous emergence of complexity in non-equilibrium systems that are nevertheless dynamically simple. Particular attention has been paid to the statistical mechanics of granular media and as a consequence, the 'sandpile' has become the touchstone of SOC. When a pile of sand is fuelled by the addition of grains it evolves into a state where, on average, the quantity of matter expelled from the pile equals the amount added to it. This dynamic equilibrium is maintained by the intermittent cascade of material down the pile through avalanches that exist on all scale-sizes up to the dimensions of the system. This sandpile paradigm has prompted workers to characterise complex systems through a simplified dynamics where fuelling is followed by reorganisation once a local critical threshold is exceeded. The reorganisation can lead to instability at other sites and consequently can establish long-scale correlated behaviour throughout the pile. The statistical description of the scale-invariant behaviour that obtains is typified by distributions possessing power law tails and in seeking a model for this, interest in the class of 'stable' or 'Lévy' distributions [2] has been rekindled.

SOC behaviour was demonstrated experimentally in the celebrated 'Oslo rice-pile' [3] and its detailed dynamics was elucidated through investigating the motion of tracer grains. The distribution for the transit times of grains through rice-piles of variable size exhibited, at longest scales, a decaying power-law extending over approximately two decades. The data show a second region of power law behaviour at the shortest time-scales for smaller sized rice-piles, although the authors did not comment upon this. A cellular automaton that purported to model this rice-pile [4] also exhibited a dual power-law at short and long scales for the distribution describing the trapping-time of grains, but again the authors confined their attention and discussion to the tail of the distribution, modelling it with Lévy statistics. Dual power-law behaviour cannot be described using Lévy statistics, which can only characterise the power-law featuring at the largest scales. One aim of this paper is to deconstruct the temporal behaviour produced by the cellular automaton and describe it in terms of a random walk model which has as key ingredient, fluctuations in the number of steps forming the walk, as in [5]. This is motivated by the well known technique [6] by which correlation may be introduced into a random walk through fluctuations in the number of steps, by which is meant that successive realisations of the walk contain different numbers of steps *independent of the properties of the step lengths.* This allows a sequence of events to be clustered (correlated) in time, for example, although their individual contributions to the quantity of interest are unrelated to the clustering process. A consequence of this is that the dual power-law behaviour for trapping times occurs naturally and, moreover, predicts the observed form for the distribution of transit times. It was also shown in [4] that the distribution of flight lengths of grains could be modelled using Lévy statistics. This raises the paradox that a spatial transport quantity can have arbitrarily large excursions, which apparently conflicts with constraints imposed by the energetics of the system. Another aim of this paper is to resolve this paradox by showing that a grain's flight is comprised of a fluctuating number of 'sub-flights' of finite length, where the number of sub-flights is described by a discrete power-law. This novel notion leads to flight lengths possessing the attributes of a Lévy distribution without the attendant problem of unphysical energetics.

Cellular automata provide an important tool for the investigation of the dynamics of sandpiles. Based on a few elementary rules, they have enabled prediction of macroscopic behaviours that can and have been observed in real systems and advanced the knowledge of the microscopic dynamics involved. However, the choice of model has not been related closely to the properties of any individual system since the objectives have, to date, been principally to illuminate the generic processes leading to SOC and their overall outcome. However, for quantitative predictions of a practical nature, it is important to understand in greater detail how the microscopic reorganisations of the system are related to the assumed numerical model. This would enable more realistic modelling of both naturally occurring and synthesised phenomena to be achieved in the future. As response to this, this paper presents a detailed statistical analysis of the internal reorganisations of a sandpile predicted by an established cellular automaton. This provides further insight into the relevance of random walk models to the subject and emphasises the importance of number fluctuations as a descriptor and motor for the dynamics.

The programme of this paper is first to review the cellular automaton used to describe the rice-pile experiment, describing in particular the physical structure of the pile in the self-organised state and how this structure influences the various behaviours of tracer grains transported through the pile. Section 3 concentrates on a spatial property, the distribution of flight-lengths and illustrates how these can be deduced from a random walk model where the number of steps in the walk fluctuates with a discrete power-law distribution. Section 4 focuses on temporal properties and presents evidence that the trapping time distribution can also be understood using a random walk model, but where now the individual steps in the walk are Lévy distributed time increments and the number of steps is a negative binomial random variate. This distribution is then used to derive the observed form of the transit time distribution of tracer grains. The final section summarises and discusses the implications of this work. Technical details pertaining to the derivations and forms of the distributions are assigned to appendices.

2 The rice-pile cellular automaton

The cellular automaton studied in [4] was designed to replicate experimental data of the Oslo rice-pile reported in [3] and has the advantage of being able to track and apply diagnostics to test particles as they move throughout the pile. The automaton therefore enables microscopic properties of particles and macroscopic attributes of the entire pile to be studied simultaneously.

The cellular automaton examines the stability of a set of slopes z_m in excess of an angle of repose, where $1 < m < L$ labels a spatial position within the pile. If the slope attains a critical gradient z_m^c at site m, 'sand' is redistributed in such a way that the gradient is reduced to a sub-critical value there and raised at nearest neighbour sites according to the rule, $z_m \geq z_m^c \Rightarrow z_{m-1} \to z_{m-1} + 1,\ z_m \to z_m - 2,\ z_{m+1} \to z_{m+1} + 1$. The critical gradient at a site is a Bernoulli random variable that fluctuates between 1 and 2 with equal probability, being reassigned whenever a grain passes over that site. Special conditions apply at the end of the pile, where the rule is modified so that if $z_L \geq z_L^c$, $z_{L-1} \to z_{L-1}+1$, $z_L \to z_L - 1$. The redistribution of grains at unstable sites occurs sequentially, counting from the left to the right of the pile, in accordance with causality. Iteration throughout the pile continues until it attains a state with $z_m < z_m^c$ everywhere. The pile is then fuelled at the top so that $z_1 \to z_1 + 1$ and the process continues. The toppling rule of the automaton used here differs slightly from that used in [4]. This does not affect the statistical properties of the tracer grains but has the advantage of being more efficient computationally and therefore allows more sophisticated diagnostics to be applied. The simplicity of the algorithm belies the complexity of behaviour it describes.

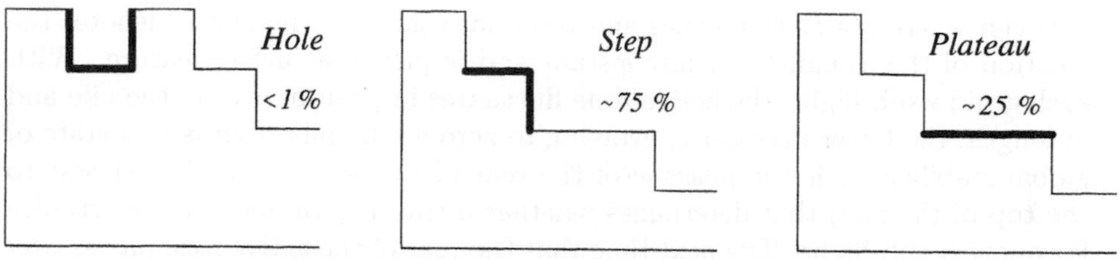

Figure 1. The essential elements from which a sandpile is comprised together with the frequency of occurrence.

Two diagnostics were used in [4] to characterise behaviour: the 'flight-length' and 'trapping-time' distributions of grains. To understand how the empirical results quoted in [4] arise and become quantifiable in terms of two distinct but simple stochastic models, novel diagnostics have to be applied to the cellular automaton that elucidate the statistical mechanics of the ensemble of grains. The way that these diagnostics inform the construction of these models is best appreciated through describing some aspects of the cellular automaton's behaviour.

A pile in the self-organised state forms a sequence of 'staircases' interspersed with 'plateaux' and, less frequently, 'holes'. These elements are illustrated in Figure 1. The algorithm predicts that a stable SOC pile has average slope $\sim 3/4$, and so $\sim 75\%$ of the piles' surface comprise sections of staircase, $\sim 25\%$ plateaux, with holes occurring with frequency $< 1\%$. The distance between consecutive plateaux is approximately exponentially distributed, with a mean inter-plateau spacing $\Delta\ell \sim 4$ grain sites and most plateaux have a length of 2 sites. Thus a stable pile of total length 400 has about 100 plateaux randomly distributed through it. The algorithm proceeds by the addition of a grain to the first site of such a pile. The stability of this site is tested and, if unstable, the grain moves to the next site, and so on until coming to rest. Grains must necessarily move down a staircase section and so the principal location at which a grain can come to rest is a plateau. Such a potential resting site has a random pre-assigned critical slope associated with it. If this is 1, the grain continues to the next potential resting site on the same or at the next plateau. If the critical slope is 2, the grain 'sticks'. In those rare instances when a grain falls into a hole, it sticks with probability 1 and forms a new plateau. Holes rapidly fill, accounting for their infrequent occurrences. Because the critical slope is either 1 or 2 with equal probability, a grain has an average sub-flight length of two inter-plateau distances, or approximately 8-grain sites before it comes to rest at a plateau. All the sites over which this grain has passed (staircases and plateaux alike) are 'perturbed' by randomly reassigning their critical slopes. Hence the potential exists for one of these sites which was initially stable to be transformed to an unstable site (and vice-versa). An unstable site will shed a grain, causing another sub-flight to ensue and continue the 'avalanche'. The order in which the stability of sites is tested always increases away from the top, effectively running down the surface of the pile. Hence there is an 'active zone' defining the spatial extent between where grains commence and terminate their sub-flights: it denotes the location of the avalanche at any instant and is purely a surface feature. With each grain's sub-flight, the active zone fluctuates in position within the pile and in length, the latter necessarily evolving to zero as the pile returns to a state of global stability. It is the position of the rear of the active zone (i.e. closest to the top of the pile) that determines whether a tracer grain has the opportunity for another sub-flight. The next time that the rear of the active zone passes over such a grain at the surface of the pile, it has the opportunity to move on another sub-flight.

An aspect to note is that the algorithm contains disparate timescales. The 'long' timescale characterises the fuelling of grains which occurs between the pile being in two consecutive stable states. The 'short' timescale characterises the redistribution of all those grains that move between fuelling events and therefore fluctuates between successive addition of grains in accordance with the size of avalanche that is produced. Despite this, the avalanche is considered to occur instantaneously on the long timescale.

3 The distribution of flight lengths

The previous section described an 'active zone', which denotes the location and spatial extent of the avalanche and occurs on the short timescale. The feature of this structure that dictates whether or not an avalanche persists is the location of the rear of the active zone m_a, for as this moves over tracer grains, its action provides them with the opportunity to move off to the next place of residence. Figure 2 shows the probability distribution for m_a for a pile of length $L = 1000$. The distribution is a power law over three decades with index -0.8. The feature appearing at small values of m_a is due to special conditions that prevail near the fuelling point. This very shallow power law implies that the active zone has no mean location. Therefore it can move anywhere over the entire pile, triggering grains to move as it does so. Indeed, this must be so, for a tracer grain would hardly ever be expelled from the pile unless the active zone could explore every part of it. Thus a property of individual grains is influenced by a macroscopic

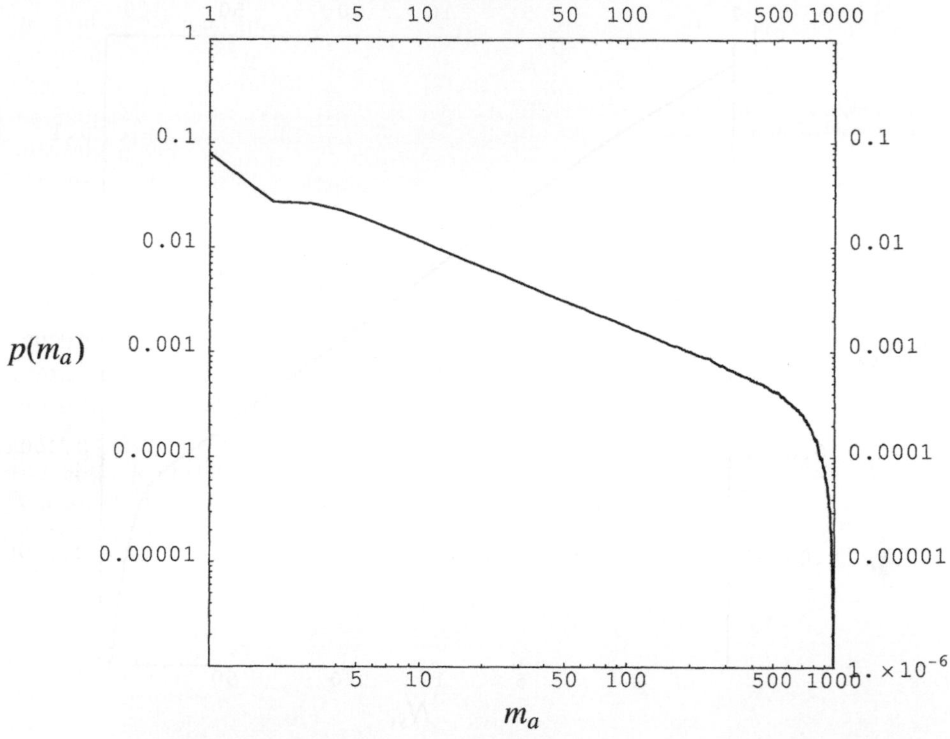

Figure 2. The probability density for the location for the rear of the active zone m_a for a pile of length $L = 1000$. The distribution is a power law with index -0.8, and is normalisable only by virtue of an end-effect. The distribution for m_a has the same power law for smaller pile lengths.

structure of the pile. A grain has the opportunity to move many times within an avalanche as m_a moves back and forth. Moreover, this spatial movement causes the excavation of interred tracer grains which affect the temporal behaviour of the pile, as will be explored in the next section.

Figure 3 shows the probability distribution for N_s, the number of sub-flights of length ℓ_i that comprise a total avalanche flight of length ℓ. The flight length $\ell = \sum_{i=1}^{N_s} \ell_i$ is the total distance travelled by a grain between fuelling events. The distribution shown is for tracer grains emanating from site 2 of the pile, principally because these have the potential for the longest flights. This discrete distribution is a power law of index -2.14. The reasons for the discrepancy between the indices of the power laws for N_s and m_a are complex. A grain does not necessarily move on another sub-flight when m_a passes over it, rather it has the *opportunity* to move. For example, the critical slope may have been reassigned to 2, or the grain may be buried below the surface, in which case the grain respectively will not, or cannot move. Moreover the distribution for N_s is

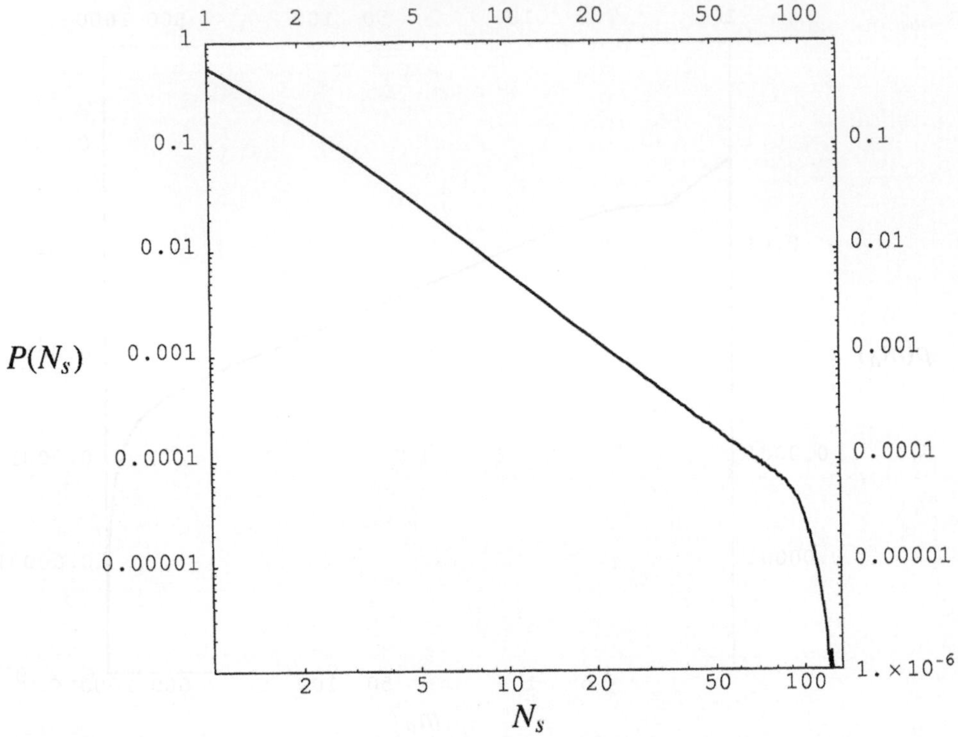

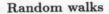

Figure 3. The distribution for the number of sub-flights N_s, which is a power law with index $\beta = -2.14$.

constructed as an ensemble average over all realisations of resultant flight lengths ℓ. There are many combinations of sub-flights ℓ_i comprising a given resultant flight ℓ and the probability for occurrence for these individual realisations would require enumeration if a match were sought. For the purpose of this paper it is sufficient to note that the underlying reason for the power law in N_s derives from the power law for m_a. It will now be shown how the above empirical observation that a tracer particle receives a power-law distributed number of 'kicks' is used to obtain the distribution for ℓ which matches the cellular automaton data. It will be shown that the resultant of the random walk is also a power law distribution with the same exponent as the number fluctuations.

In section 2 the motion of grains was described as being a series of 'jumps' or sub-flights which have an average length of 8 sites, being 2 inter-plateau lengths. The actual distribution of inter-plateau lengths matters little to what follows and can be taken as constant. The sub-flights can then be considered as an integer multiple of a fixed inter-plateau length $\Delta\ell$ so that $\ell_n = n\Delta\ell$. Suppose that the probability of traversing any possible resting site is $1/\mu$ where $\mu > 1$. For the cellular automaton $\mu = 2$ because the critical slope at the site of rest adopts one of two values. The probability of the grain travelling over $n-1$ independent potential resting sites before remaining at the n-th site is therefore $(\mu)^{n-1}(1-1/\mu)$. Hence the generating function for obtaining a sub-flight of length ℓ_n is

$$q(s) = \langle \exp(-s\ell_n) \rangle = (\mu-1)\sum_{n=1}^{\infty}\left(\frac{1}{\mu}\right)^n \exp(-ns\Delta\ell) = \frac{\mu-1}{\mu\exp(s\Delta\ell)-1}.$$

The form of the generating function near $s=0$ determines the large-scale asymptotic behaviour, i.e.

$$q(s) \sim \frac{1}{1+\mu s\Delta\ell/(\mu-1)}$$

which is the generating function of the exponential distribution. A random walk comprising N such independent sub-flights has resultant

$$\ell = \sum_{n=1}^{N} \ell_n \tag{3.1}$$

with generating function $q_N(s) = q(s)^N$, and this generates the Gamma distribution:

$$p_N(\ell) = \left(\frac{(\mu-1)\ell}{\mu\Delta\ell}\right)^{N-1} \frac{(\mu-1)}{\mu\Delta\ell\Gamma(N)} \exp\left(-\frac{(\mu-1)\ell}{\mu\Delta\ell}\right),$$

all moments of which exist. The model adopted for the number fluctuations that is consistent with Figure 3 is taken to be

$$P(N) = \frac{1}{\zeta(\beta)N^{\beta}}, \qquad N \geq 1, \quad \beta > 1 \tag{3.2}$$

where the Riemann Zeta function $\zeta(\beta)$ [7] provides normalisation. Because the number of steps is independent of the length of each step, the distribution for the resultant flight length ℓ that results from averaging over all realisations of N is:

$$\begin{aligned} p(\ell) &= \sum_{N=1}^{\infty} P(N)p_N(\ell) \\ &= \frac{(\mu-1)\exp\left(-(\mu-1)\ell/\mu\Delta\ell\right)}{\mu\Delta\ell\zeta(\beta)} \sum_{N=0}^{\infty}\left(\frac{(\mu-1)\ell}{\mu\Delta\ell}\right)^N \frac{1}{N!(1+N)^{\beta}}. \end{aligned}$$

Setting $x = (\mu-1)\ell/\mu\Delta\ell$ obtains

$$p(x) = \frac{\exp(-x)}{\zeta(\beta)} \sum_{N=0}^{\infty} \frac{x^N}{N!(1+N)^{\beta}}. \tag{3.3}$$

An alternative expression for this pdf can be obtained on using the definition of the Gamma function [7] to write

$$(1+N)^{-\beta} = \frac{1}{\Gamma(\beta)} \int_0^\infty du\, u^{\beta-1} \exp\left(-(1+N)\right)$$

whereupon the summation can be evaluated to obtain the equivalent integral representation for the pdf (3.3):

$$p(x) = \frac{\exp(-x)}{\zeta(\beta)\Gamma(\beta)} \int_0^\infty du\, u^{\beta-1} \exp(-u) \exp\left(x \exp(-\mu)\right). \tag{3.4}$$

This distribution, expressed either in the form (3.3) or (3.4) constitutes the first principal result of this paper.

The behaviour of (3.4) is still not particularly evident, but can be made transparent by writing the 'exponential of an exponential' as [8, 5]:

$$\exp\left(x \exp(-u)\right) \approx 1 + (\exp(x) - 1) \exp\left(-\frac{xu}{(1-\exp(-x))}\right),$$

which has the advantage of reducing to the conventional steepest descents approximation for large x and is correct for arbitrary values of x if u is sufficiently small. Use of this approximation therefore yields the correct asymptotic behaviour for large and small values of x. The remaining integrals are straightforward to perform and give:

$$p(x) \sim \frac{\exp(-x)}{\zeta(\beta)} + \frac{(1-\exp(-x))^{1+\beta}}{\zeta(\beta)\,(1-\exp(-x)+x)^{\beta}}$$

as an approximation for the pdf (3.3-3.4) which conveniently reveals its structure. If x is small, $p(x) \sim 1/\zeta(\beta)$, whereas if x is large, the distribution has a power-law tail with $p(x) \sim 1/x^{\beta}$. In appendix 1 it is shown that the asymptotic form of the tail can be determined with greater precision to be:

$$p(x) \sim \frac{1}{\zeta(\beta)x^{\beta}} \left(1 + \frac{3 + 4(\beta+1/2)(\beta-3/2)}{8x} + \ldots\right) \tag{3.5}$$

when $x \gg 1$. The r-th moment of (3.3-4) exists only if $r < \beta - 1$, a property inherited from the parent distribution (3.1). In particular a mean flight length $\langle \ell \rangle$ exists if $\beta > 2$ and an elementary calculation obtains

$$\langle \ell \rangle = 2\Delta\ell \frac{\zeta(\beta-1)}{\zeta(\beta)} \tag{3.6}$$

and so a grain moving through a pile of total length L has on average $M = L/\langle \ell \rangle$ flights before leaving the system.

The distinction between the distribution describing this random walk and a Lévy distribution is important to clarify, even though they ostensibly appear to

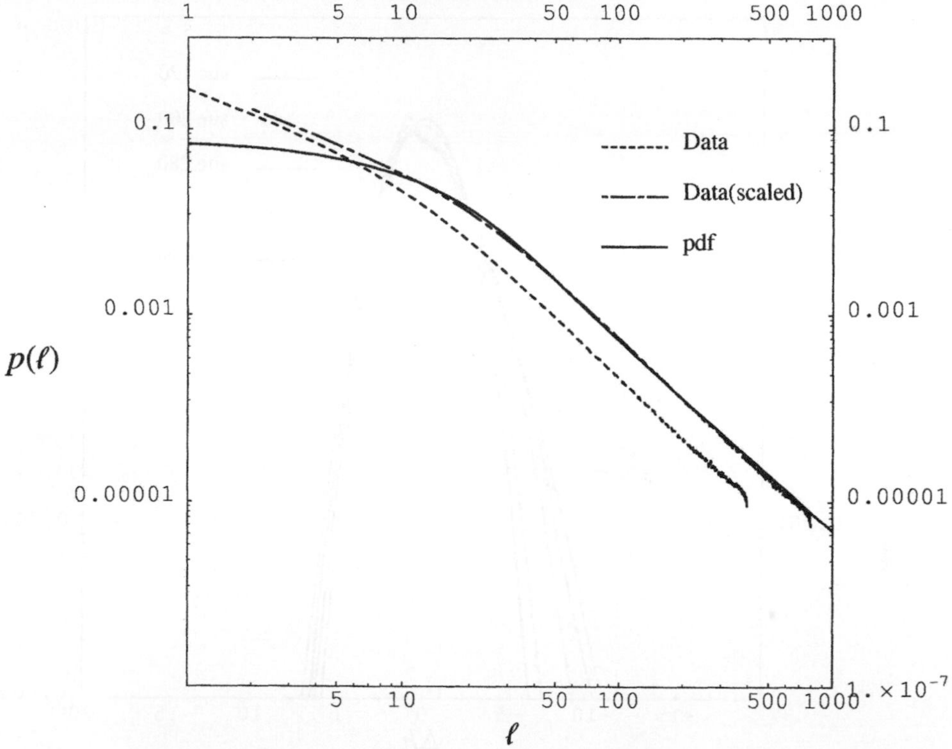

Figure 4. The dashed curve is the distribution of flight-lengths obtained from the cellular automaton for a pile of length $L = 400$. The linearly scaled data is shown by the chain curve and is compared with the distribution given by equation (3.3) shown by the full curve.

have similar asymptotic forms. The power-law tail in a Lévy random walk occurs as a consequence of the power-law distributed individual step lengths. Lévy random walks are inappropriate for describing, for example, spatial movement of material where the energetics would prohibit the occurrence of arbitrary sized step lengths. In such instances it is more appropriate for the individual steps to have finite integer moments, but with the power law behaviour for the resultant ultimately deriving from another mechanism such as number fluctuations. Moreover, the index of the power-law appearing in (3.2) is not restricted to lie in the range for that of the stable distributions.

The dashed curve in figure 4 shows the distribution of ℓ obtained from the cellular automaton for all grains emanating from site 2 of the pile. This has a power law tail with index -2.14 and can be readily explained using the information contained in figure 3 together with the random walk described above. The full curve shows the distribution (3.3) where $\Delta\ell = 4$ has been used, in accord with the value deduced from the average slope of a SOC pile. The chain curve

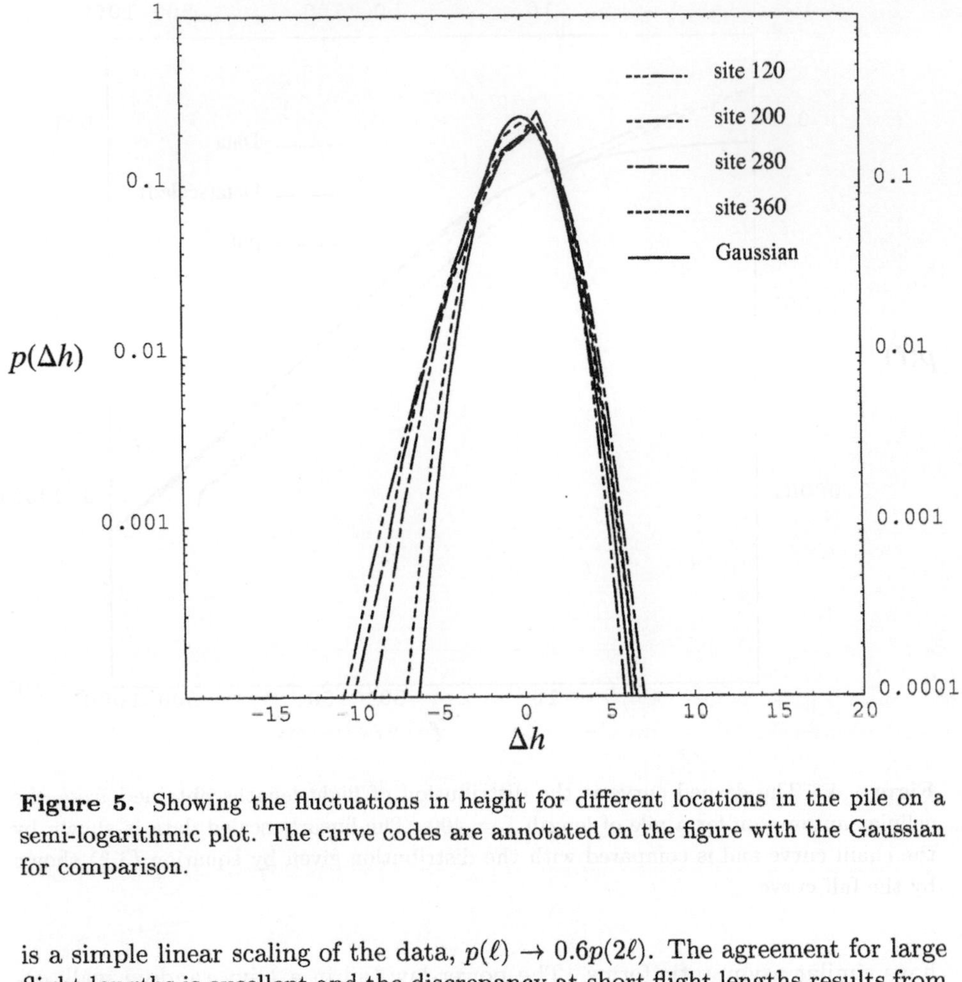

Figure 5. Showing the fluctuations in height for different locations in the pile on a semi-logarithmic plot. The curve codes are annotated on the figure with the Gaussian for comparison.

is a simple linear scaling of the data, $p(\ell) \to 0.6p(2\ell)$. The agreement for large flight lengths is excellent and the discrepancy at short flight lengths results from the simplifying assumption of uniform inter-plateau distances that was used to derive (3.3).

4 The distribution of trapping times and transit times

The second diagnostic used in [4] was the distribution of 'trapping-times', which is the time a grain remains at rest at a particular site. The trapping time occurs on the long timescale that is characterised by the rate of fuelling. Each site is monitored to give a distribution of trapping-times for the pile as a whole and it was found that this distribution has a power-law tail and other features which the following model can fully account for.

A grain has the opportunity to move, and thereby end a trapping-time period, only if it is on the surface. Grains that are buried must wait until they are

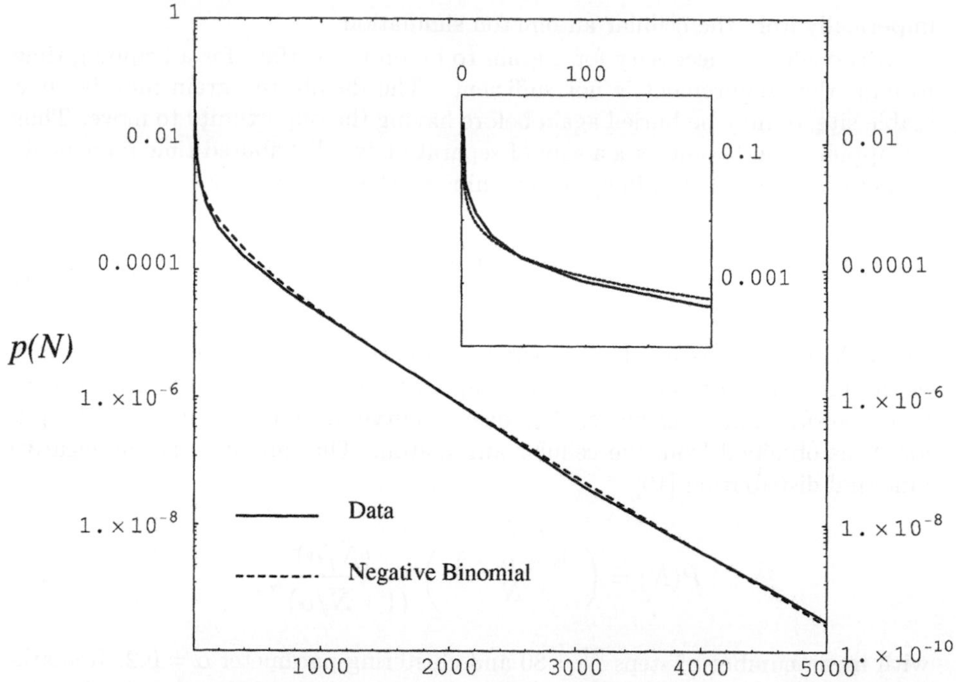

Figure 6. The dashed curve shows the probability density for the number of instances that a grain is exposed at the surface of the pile but does not move, derived from cellular automaton data. The full curve is the negative binomial distribution with $\overline{N} = 80, \alpha = 0.2$. Both are shown on semi-logarithmic plots. The inset resolves the pdf for smaller numbers of exposures.

excavated. Figure 5 shows the distribution for increments in height Δh of the pile at different locations removed from the central fuelling point. The increments are essentially stationary, have zero mean but are skewed and so deviate from a Gaussian distribution. Supposing for the moment the distribution of height fluctuations were approximated by a Gaussian distribution, the instances when a particular grain returns to the surface may then be interpreted as the first return time of a Brownian fractal. The distribution for such a return time has a power law tail [9]. In fact, because the height fluctuations are distinct from a Gaussian leads to being able to quantitatively determine the index of the power-law tail. However the execution of this calculation itself requires a number of technical innovations which have a currency beyond applications to sandpiles, and so the details are presented elsewhere [12]. For the purposes of this paper it is entirely correct to model the time t_m for a grain to return to the surface with a Lévy distributed random variable. The index of the distribution used will be obtained imperically from the cellular automaton simulation.

Although it is necessary for a grain to be on the surface for a trapping time to end, this requirement is not sufficient. The disinterred grain may be at a stable site, or may be buried again before having the opportunity to move. Thus a trapping time t comprises a sum of separate Lévy distributed time increments t_m between a grain remaining or returning to the surface, viz.:

$$t = \sum_{m=1}^{N} t_m \tag{4.1}$$

where N denotes the number of instances that a grain is exposed at the surface of the pile but does *not* move. The trapping time ends when the grain comes to the surface and *does* move. The dashed curve in figure 5 illustrates the pdf for N as obtained from the cellular automaton. The full curve is the negative binomial distribution [10]

$$P(N) = \binom{N+\alpha-1}{N} \frac{\left(\overline{N}/\alpha\right)^N}{\left(1+\overline{N}/\alpha\right)^{N+\alpha}} \tag{4.2}$$

with mean number of steps $\overline{N} = 80$ and clustering parameter $\alpha = 0.2$. Remarkably, the data are indistinguishable from this model over 3-decades of values for N and the inset shows that the model is accurate at very small values of N also. The reason why the number of re-surfaces/re-burials of a tracer grain at the surface of the pile should be accurately described by the negative binomial distribution is unclear but the empirical evidence is compelling. The distribution (4.2) has two parameters, the mean $\overline{N}$ and $\alpha > 0$ is the cluster parameter. The special case $\alpha = 1$ is the Bose-Einstein or geometrical distribution, which describes 'thermal' number fluctuations and when $\alpha \to \infty$ it is the Poisson distribution, describing purely random number of steps. The smaller the value of α, the greater is the strength of clustering. Insofar as the statistical description of clustering is an inherent feature of the distribution, it again is appropriate to use it in a variable step-number random walk to incorporate the effect of correlations. Bunching arises in the pile in the following qualitative fashion. Consider a system-wide avalanche in which a large number of tracer particles move from, or re-surface at a particular site in the pile. Grains that have moved come to rest elsewhere and therefore commence a new trapping time. Those that have resurfaced end their present time and commence a new one. After such an avalanche, successive fuelling events tend to result in small sized avalanches confined to the top of the pile as the slope increases from a sub-critical state. With each feed, all surface tracer grains at locations further down the pile receive identical time increments. Eventually the avalanches become larger and reach groups of surface tracer grains, bringing to an end several trapping times simultaneously. Many of these trapping times will comprise a similar number of time increments which provides the clustering in N.

The above picture will now be developed into a random walk model for the distribution of trapping times. The ingredients are individual 'sub-trapping time' increments t_m, which are power-law distributed and can therefore be modelled by a Lévy distribution, together with a discrete number N of times that a grain comes to the surface of the pile without moving from that site, the statistics for which is described by (4.2). This model is similar to that introduced in [5]. The difference here is that the time increments t_m are all positive, so that the theory must be recast in terms of the one-sided Lévy distributions to model t_m.

In appendix 2 it is shown that on using a one-sided Lévy distribution to describe the time increments with number fluctuations described by (4.2), obtains the following limit distribution for the trapping time on letting the average number of steps $\overline{N} \to \infty$:

$$p_{\alpha,\gamma}(t) = \mathrm{Re}\frac{1}{\pi}\int_0^\infty (1 + u^\gamma(1 + i\Lambda)/\alpha)^{-\alpha} \exp(iut)\, du \tag{4.3}$$

where:

$$\Lambda = \begin{cases} \frac{2}{\pi}\ln|u| & \gamma = 1 \\ \tan(\gamma\pi/2) & \gamma \neq 1 \end{cases}$$

This density function is valid if the individual steps in the walk are drawn from the class of stable Lévy processes or from distributions having power-law tails *similar* to Lévy distributions and it constitutes the second principal result of this paper. In common with its parent Lévy distribution, the two parameter distribution (4.3) has divergent integer moments. Although not of the stable class, this distribution is infinitely divisible and this property has implications for explaining some features in data that follows. For the special case $\gamma = 2$, the distribution is that of a Gamma-variate.

The asymptotic behaviour of the distribution (4.3) can be deduced for different values of α and γ. When $t \gg 1$ the form adopted by (4.3) is identical to that of the Lévy distribution with the same index γ, i.e. $p_{\alpha,\gamma}(t) \sim t^{-\gamma-1}$ giving scale invariant behaviour in this regime. Specifically:

$$p_{\alpha,\gamma}(t) \sim \frac{2\Gamma(1+\gamma)}{\pi}\sin\left(\frac{\pi\gamma}{2}\right)t^{-1-\gamma} \qquad 0 < \gamma < 2, \quad x \gg 1. \tag{4.4}$$

When $t \ll 1$, the form of the distribution depends upon the size of both γ and the product $\alpha\gamma$ when compared with unity. The important property to note is that the distribution can have either an increasing or decreasing power-law at small values of t for particular values of t the parameters α and γ. Specifically:

$$
p_{\alpha,\gamma}(t) \sim \begin{cases}
\dfrac{(\alpha|\cos(\gamma\pi/2)|)^{\alpha}}{\gamma\Gamma(\alpha,\gamma)} \dfrac{\sin(\alpha\pi(\gamma-1))}{\sin(\alpha\pi\gamma)} t^{\alpha\gamma-1} & \alpha\gamma < 1, \gamma > 1 \quad 4.5(a) \\
\dfrac{(\alpha|\cos(\gamma\pi/2)|)^{1/\gamma}\Gamma(1+1/\gamma)\Gamma((\alpha\gamma-1)/\gamma)\sin(\pi/\gamma)}{\pi\Gamma(\alpha)} & \alpha\gamma > 1, \gamma > 1 \quad 4.5(b) \\
\dfrac{(\alpha|\cos(\gamma\pi/2)|)^{\alpha}}{\Gamma(\alpha,\gamma)} t^{\alpha\gamma-1} & \text{all } \alpha\gamma, \gamma < 1 \quad 4.5(c) \\
\dfrac{1}{2}\left(\dfrac{\cos(\gamma\pi/2)}{\gamma}\right)^{1/\gamma} & \alpha\gamma = 1, \gamma < 1 \quad 4.5(d) \\
-\dfrac{\sin(\pi/\gamma)}{\pi}\left(\dfrac{|\cos(\gamma\pi/2)|}{\gamma}\right)^{1/\gamma} \ln(x) & \alpha\gamma = 1, \gamma > 1 \quad 4.5(e)
\end{cases}
$$

Attention is drawn to the behaviour of this distribution in those instances when $\gamma > 1$. If $\gamma > 1$ and $\alpha\gamma < 1$, the distribution possesses two separate power-law behaviours, in the tail with $p_{\alpha,\gamma}(t) \sim t^{-\gamma-1}$ and at small values of t where $p(t) \sim t^{\alpha\gamma-1}$, (4.5b). If $\alpha\gamma > 1$ the distribution retains its power-law tail but has an inner-scale (4.5d).

The effect of finite $\overline{N}$ is to modify the distribution, so that the pdf is now:

$$
p_{\overline{N}}(t) = \mathrm{Re}\frac{1}{\pi}\int \left[1 + \frac{\overline{N}}{\alpha}\left(1 - \exp\left(-u^{\gamma}(1+i\Lambda)/\overline{N}\right)\right)\right]^{-\alpha} \exp(iut)du \qquad (4.5)
$$

which has three parameters, $\overline{N}, \gamma$ and α. Analysis identical to that shown in [5] reveals that $\overline{N}$ introduces an inner-scale which resolves the inner power-law that occurs when $\alpha\gamma < 1$. This inner-scale extends out to $t \sim \gamma(\overline{N})^{-1/\gamma}$. Beyond this region, the pdf matches onto the inner power-law of the limit distribution whereon the two distributions become indistinguishable.

Figure 7 shows the trapping time distribution obtained from the cellular automaton, which has a power law tail of index ν_t, so that $\gamma = -1.16$ indicated by the dashed curve. This, together with the information derived from figure 5, $\overline{N} = 80, \alpha = 0.2$ obtains all the parameters required to apply the distributions (4.3) and (4.5). For $\overline{N} \sim 80$ the inner-scale occurs at $t \sim 0.027$.

Figure 3 of [4] shows the transit-time distribution for grains passing through the entire pile. This has the same tail as the distribution for trapping-times but is constant for small transit-times, i.e. it exhibits an inner-scale for short transit-times. This behaviour is readily understood in terms of the model presented here since the transit-time is the independent sum of individual trapping-times. The characteristic function of the distribution of the transit-times is of the same form as that in (4.3) by virtue of infinite divisibility, but with different parameter, viz.:

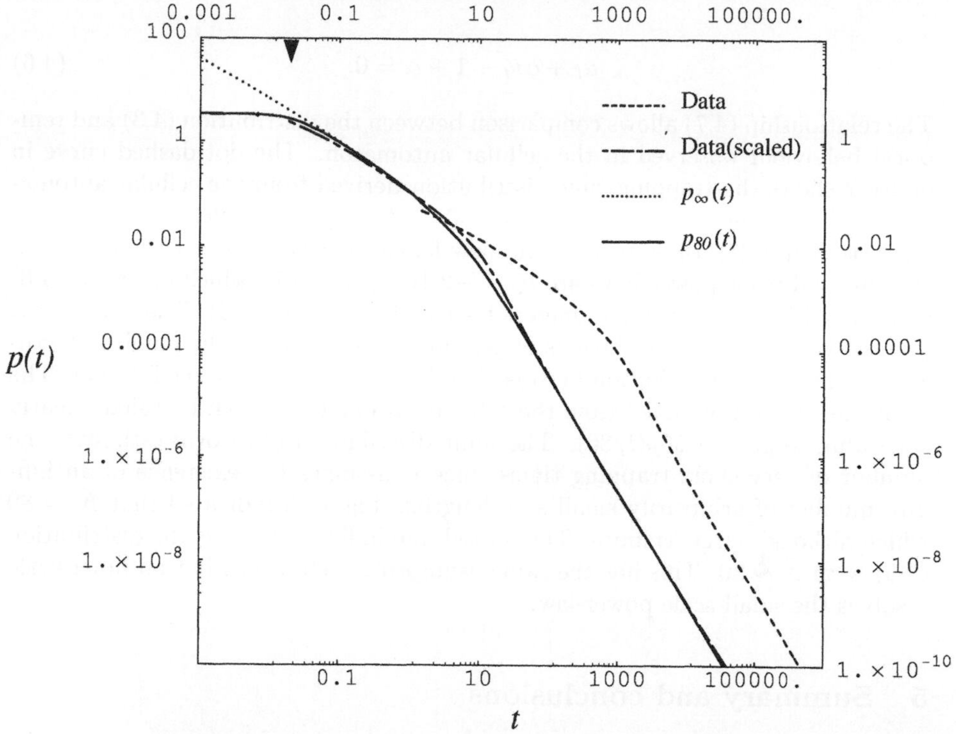

Figure 7. The dashed curve shows the trapping time distribution for the cellular automaton which is then linearly scaled given by the chain curve for comparison with the limit distribution (2.2) shown in full. The curve shows the pdf for finite $\overline{N}$ which introduces an inner-scale indicated by the thorn on the upper axis. The values of $\alpha = 0.2$ and $\overline{N} = 80$ are those derived from the pdf illustrated in figure 6.

$$\left(1 + u^{\gamma}(1 + i\Lambda)/\alpha\right)^{-\overline{M}\alpha},$$

with $\overline{M}$ the average number of trappings on the long timescale that a grain experiences before being expelled from the pile. This distribution has the same power-law asymptotic form for large values of t as (4.4). The behaviour for small values of t depends on the value of $\overline{M}$ which can be estimated from (3.6). $\overline{M}$ is greater than unity for systems of size $L > 40$ but $\overline{M}\alpha\gamma > 1$ for system sizes $L > 175$. Thus for system sizes greater than ~ 175, the transit time distribution will exhibit an inner-scale.

The asymptotes (4.4) and (4.5 a) that predict the power-law indices ν_t and ν_f associated with the tail and 'front' of the distribution respectively are connected through the cluster parameter by:

$$\nu_f + \alpha\nu_t + 1 + \alpha = 0. \tag{4.6}$$

The relationship (4.7) allows comparison between the distribution (4.3) and temporal behaviour observed in the cellular automaton. The dot-dashed curve in figure 7 shows the trapping-time distribution derived from the cellular automaton [4] for a sandpile of length $L = 400$. Although sandpiles with $L > 400$ were studied in [4], these revealed no new features. From figure 7 the indices of the tail and front power laws are $\nu_t = -2.16, \nu_f = -0.78$ which on using (4.6) gives $\gamma = 1.16, \alpha = 0.2$ as parameters for the distribution (4.3). The value for α accords with the value of the cluster parameter obtained for the number fluctuations. The limit distribution (4.3) is also shown in figure 7 by the full line. The chain line is data obtained from the cellular automaton, which is scaled linearly according to $p(t) \to 50p(t/80)$. The limit distribution (4.3) over-estimates the number of very short trapping times since it assumes the existence of an infinite number of arbitrarily small step lengths. Figure 6 indicated that $\overline{N} \sim 80$ which although large, is finite. The dotted line in figure 7 shows the distribution (4.5) with $\overline{N} \sim 80$. This has the same asymptotic behaviour, but an inner scale resolves the small-scale power-law.

5 Summary and conclusions

This paper has applied novel diagnostics to a rice-pile simulation. These diagnostics have indicated that the spatial and temporal behaviours of tracer grains can be described in terms of random walk models. Crucially these random walks require the incorporation of step number fluctuations into their formulation. The first model describes the distribution for the resultant flight lengths of grains. This is shown to comprise a sum of relatively short sub-flights whose number fluctuates according to a discrete power-law. The index of the power-law describing the resultant flight lengths is inherited from the number fluctuation distribution. The second model describes a temporal quantity, the distribution of times that a grain remains at rest before being transported to another site by an avalanche. The mechanism at work here is more subtle and relies on the height of the pile at a particular site fluctuating and thereby bringing a tracer grain to the surface at particular times. The distribution of times between such returns to the surface can be modelled with a one-sided Lévy distribution. The number of times a grain returns to the surface before being transported to another site is described accurately by a negative binomial distribution and so a trapping time comprises a sum of Lévy distributed step lengths which fluctuate in number. The power-law tail of the trapping time distribution is inherited from the individual Lévy steps which correspond to the distribution of those times when a grain is buried and so cannot move. The index for the tail of the distribution is intimately related to the form of the height fluctuations and can be quantitatively determined, however this calculation itself requires distinct innovations and so the details are

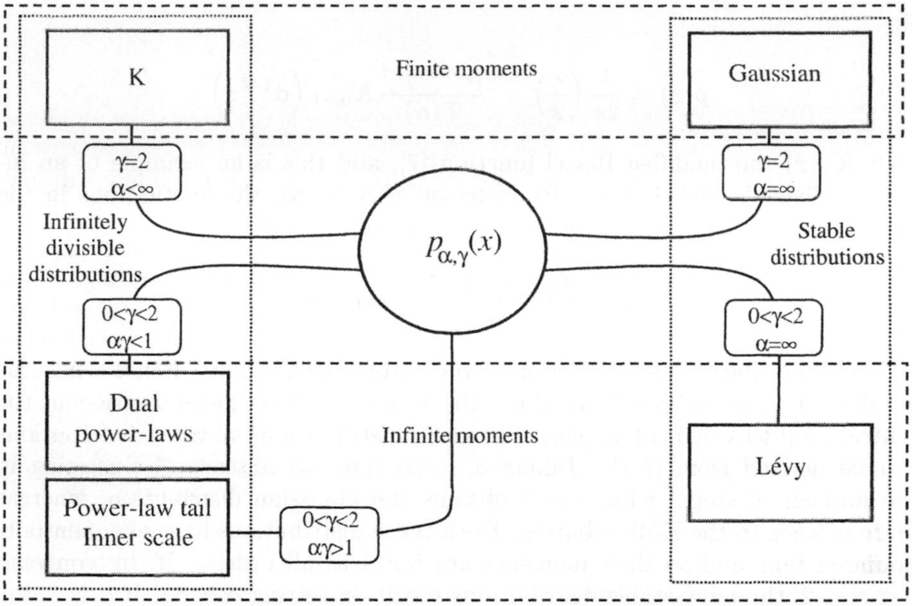

Figure 8. The relationship of the two-sided distribution (5.1) to other known distributions. The forms adopted by $P_{\alpha,\gamma}$ appearing in the lower left-hand corner of this 'phase diagram' are related to the behaviours produced by the sand-pile cellular automaton.

presented elsewhere [12]. Features of the distribution occurring at small times are governed by the clustering introduced by the number fluctuations.

The distribution (4.3) was formulated to specifically account for trapping times, which naturally requires a forward-going distribution for each of the steps. Relaxing this constraint to the situation where the increments can be positive or negative illustrates the more intimate connections with the Gaussian and other stable distributions and to the K-distribution which is a generic distribution occurring in the coherent scattering of radiation from random media. If the individual steps in the walk are stable-distributed, the distribution (4.3) simplifies to:

$$p(x) = \frac{1}{2\pi}\int_{-\infty}^{\infty}\left(1+\frac{|u|^{\gamma}}{\alpha}\right)^{-\alpha}\exp(iux)\,du \tag{5.1}$$

in the large $\overline{N}$ limit. This distribution shares most of the properties of the distribution (4.3) having a Lévy tail with $p(x) \sim x^{-1-\gamma}$ irrespective of the size of α, a power-law 'front' with $p(x) \sim x^{\alpha\gamma-1}$ for $x \ll 1$ if $\alpha\gamma < 1$ and an 'inner-scale' with $p(x) \sim$ const. if $\alpha\gamma > 1$. For the marginal case with $\alpha\gamma = 1$, $p(x) \sim -\ln(x)$. Again, it is not possible to write down explicit forms for (5.1) in terms of tabulated functions, but certain limits can de deduced for specific parameter regimes. If $\gamma = 2$, the step lengths are Gaussian, and x is exactly K-distributed

[6]:

$$p(x) = \frac{1}{2\pi}\left(\frac{x}{2}\right)^{\alpha-1}\frac{\alpha^{\alpha/2+1/4}}{\Gamma(\alpha)}K_{\alpha-1}\left(\alpha^{1/2}x\right)$$

with $K_\alpha(x)$ the modified Bessel function [7], and this is an example of an infinitely divisible distribution. By contrast, if $\alpha \to \infty$, the fluctuations in the number of steps comprising the walk are Poissonian and

$$\lim_{\alpha\to\infty}\left(1+\frac{u^\gamma}{\alpha}\right)^{-\alpha} = \exp(-u^\gamma)$$

which is the characteristic function of the stable distributions. Hence when the number of steps are purely random, the Lévy-Gnedenko generalisation of the central limit theorem [13] applies. Figure 8 illustrates how all these distributions can be derived from (5.1). Taking $\alpha = \infty$ (i.e. an absence of clustering in the number of steps) with $\gamma = 2$ obtains the Gaussian distribution, whereas finite α leads to the K-distribution. Both these distributions have exponentially bounded tails and so their moments are finite to all orders. If, by contrast, $0 < \gamma < 2$, the Lévy-stable distributions result on setting $\alpha = \infty$. If α is finite, then a distribution with Lévy tail and a finite inner-scale results if $\alpha\gamma > 1$, and dual power-laws if $\alpha\gamma < 1$. Both these are examples of infinitely divisible distributions, but like the Lévy-stable distributions to which they are related, the variance and higher moments do not exist.

There are several routes via which this work can be extended and exploited. A technical study of the properties of clustered Lévy random walks formed in higher dimensions and from a finite number of steps is presented elsewhere [5]. The random walk with power-law distributed number of steps is novel and warrants further investigation. For example, the value of the index β will affect the rate of convergence to either Gaussian or other stable distributions. The properties of such random walks either on the line or in higher dimensions is of relevance to the study of macroscopic transport phenomena. Analysing the flights made by tracer particles down SOC profiles can provide a paradigm for motion through unstable or turbulent media and thereby elucidate aspects of anomalous transport in complex systems. At a deeper level, the stochastic processes which create *discrete* power distributions are of an intrinsic and fundamental interest. The mechanism by which power-law behaviour is manifested in temporal properties of the pile has been linked to the classical problem of 'first return' or 'first passage' of a stochastic process.

Earlier work produced by a large collective of authors has shown how simple cellular automata can reproduce some of the effects observed in the evolution of sandpiles. Such work has helped to identify the microscopic mechanisms that may play an important but hidden rôle in determining the co-operative behaviour of complex systems. Adopting a principally computational approach has not, however, provided much new physical insight into the processes taking place. This paper has performed a detailed analysis of the statistics of both the microscopic transitions and the macroscopic changes in a sandpile. In so doing it has revealed patterns of behaviour which are amenable to physical interpretation and for which stochastic models either exist or can be developed. With regard to these developments, the discrete power-law distribution provides a potentially fruitful vein for further investigation. The construction of population models describing processes whose equilibrium or limiting forms have these unusual statistics warrants further study since these processes evidently underpin the dynamics of SOC behaviour at the deepest level.

Acknowledgements

It is a pleasure to acknowledge Dr.Á. Corral's clarifications of some details in reference [4]. This work is supported by the UK Engineering and Physical Science Research Council and the Leverhulme Trust.

Appendix 1

This appendix contains an asymptotic analysis of the spatial random walk for the flight lengths. This requires evaluation of the pdf (3.3):

$$p(x) = \frac{\exp(-x)}{\zeta(\beta)} \sum_{N=0}^{\infty} \frac{x^N}{N!(1+N)^\beta}$$

which, with the aid of Stirling's formula to represent $N!$, can be approximated by the integral:

$$p(x) = \frac{1}{\zeta(\beta)(2\pi)^{1/2}} \int_0^\infty \frac{\exp\left(y\ln(x) + y - y\ln(y) - x\right) dy}{y^{1/2}(1+y)^\beta}.$$

The argument of the exponential function has a single turning point at $y = x$, where the numerator of the integrand attains a maximum value of unity. The value of the integral will therefore be dominated by behaviour of the integrand in the vicinity of $y = x$. Expanding the argument of the exponential function to second order about this point leads to the approximation:

$$\exp\left(-(y-x)^2/2y\right)$$

which possesses the correct behaviour near $y = x$ when x is large. Setting and considering $x \gg 1$ obtains:

$$p(x) = \frac{x^{1/2-\beta}}{\zeta(\beta)(2\pi)^{1/2}} \int_{-1}^\infty \frac{\exp\left(-xu^2/\left(2(1+u)\right)\right) du}{(1+u)^{\beta+1/2}}.$$

This integral is in a form that is amenable to the analysis given in [11], whereupon the first two terms in the expansion are readily found to be those given by (3.5) in the text.

It is simple to show that for small values of x,

$$p(x) \sim \frac{1}{\zeta(\beta)}\left(1 - x\left(1 - \frac{1}{2^\beta}\right) + \frac{x^2}{2}\left(1 + \frac{1}{3^\beta} - \frac{1}{2^{\beta-1}}\right) + \ldots\right).$$

Appendix 2

This appendix derives the distribution (4.3) which describes fluctuations in the random variable

$$t = \sum_{m=1}^{N} t_m$$

where the t_m are independent but statistically identical one-sided Lévy distributed random variables with characteristic function

$$C_L(u) = \exp\left(-|u|^\gamma \left(1 + i\mathrm{sgn}(u)\Lambda\right)\right),$$

and N fluctuates according to the negative binomial distribution (4.2). The average characteristic function that results from considering all realisations of N is

$$\begin{aligned} C(u) &= \sum_{N=0}^{\infty} P(N) C_L(u)^N \\ &= \left(1 + \frac{\overline{N}}{\alpha}(1 - C_L(u))\right)^{-\alpha} \end{aligned}$$

and the distribution for $p(t)$ follows on Fourier transforming this. Noting that $C_L(-u) = C_L(u)^*$ obtains

$$p(t) = \mathrm{Re}\frac{1}{\pi}\int_0^\infty du \exp(iut) C(u).$$

On rescaling u through $u \to u/\overline{N}^{1/\gamma}$ followed by a scaling in $t \to t\overline{N}^{1/\gamma}$ gives the distribution (4.6), whence upon letting $\overline{N} \to \infty$ obtains the limit distribution (4.3).

Bibliography

1. P. Bak, C. Tang & K. Wiesenfeld, Phys. Rev. Lett. **59**, 381, (1987)., P. Bak, C. Tang & K. Wiesenfeld, Phys. Rev. A, **38**, 364, (1988). E.T. Lu & R.J. Hamilton, Astrophys J. **380**, L89, (1991). K. Christensen, S.R. Nagel, Rev. Mod. Phys. **64**, 321, (1992). S. Mineshige, M. Takeuchi & H. Nishimori, Astrophys. J. **435**, L125, (1994). H.L. Swinney, Physica D, **76**, 70, (1994). D. H. Zanette & P. A. Alemany, Phys. Rev. Lett. **75**, 366, (1995). R.D. Pinto, W.M. Gonçalves, J.C. Sartorelli, & P.M.C. de Oliveira, Phys. Rev. E. **52**, 6896, (1995). B.M. Boghosian, Phys. Rev. E. **54**, 4754, (1996). R.O. Dendy & P. Helander, Plasma Phys. Control Fusion **39**, 1947, (1997). A. Chessa, H.E. Stanley, A. Vespignani & S. Zaperi, Phys. Rev. E, **59**, R12,(1999). P. Bak, *How Nature Works*, (Oxford University Press, Oxford, 1997).

2. P. Lévy, *Théorie de l'Addition des Variables Aléatories*, (Gauthier-Villars, Paris, 1937).

3. K. Christensen, Á. Corral, V. Frette, J. Feder & T. Jøssang, Phys. Rev. Lett. **77**, 107, (1996). V. Frette, K. Christensen, A. Mathe-Sørrenssen, J. Feder ,T. Jøssang & P. Meakin, *Nature*, **379**, 49,(1996).

4. M. Boguñá & Á. Corral, *Phys. Rev. Lett.* **78**, 4950, (1997).

5. K.I. Hopcraft, E. Jakeman & R.M.J. Tanner, *Phys. Rev. E.*, **60**, 5327,(1999).

6. E. Jakeman & P.N. Pusey, *IEEE Trans. Antennas Propag.* **AP24**, 806, (1977).E. Jakeman & P.N. Pusey, *Radar 77* (IEE, London), p 105. E. Jakeman & P.N. Pusey, Phys. Rev. Lett. **40**, 546, (1978). E. Jakeman, *J. Phys.* A: Math. Gen. **13**, 31, (1980). E. Jakeman & R.J.A. Tough, *Adv. Phys.* **37**,471,(1988). E. Jakeman, *J. Opt. A. Pure Appl. Opt.* **1**, 784,(1999).

7. M. Abramowitz & I.A. Stegun, *Handbook of Mathematical Functions*, 9th Ed. (Dover, New York, 1970).

8. M.V. Berry, Philos. Trans. R. Soc. London, Ser. A **273**, 611, (1973).

9. W. Feller, *An Introduction to Probability and Its Applications*, Vols I & II. 2nd Edition, (John Wiley & Sons, London & New York, 1971). M. Ding & W. Yang, *Phys. Rev. E*, **52**, 207, (1995).

10. M. Evans, N. Hastings & B. Peacock, *Statistical Distributions*, 2nd Ed. (Wiley, New York, 1993).

11. R.B. Dingle, *Asymptotic Expansions: their Derivation and Interpretation.* (Academic, London & New York, 1973).

12. K.I. Hopcraft, R.M.J. Tanner, E. Jakeman, J.P. Graves, 'Fractional non-Brownian motion and the trapping-times of grains in a rice-pile', submitted to Physical Review E, January 2001.

13. G. Christoph & W. Wolf, *Convergence Theorems with a Stable Limit Law* (Akademie, Berlin, 1992), p. 17.

Fractional Integrals, Singular Measures and Epsilon Functions

R.F. Hoskins

ISS, SERC, Hawthorn Building, De Montfort University, Leicester LE1 9BH, UK

Abstract

This paper examines the speculative theory advanced by N.R. Nigmatullin on the possible relations between fractal curves and fractional integration operators. In order to clarify the discussion it has been found convenient to introduce an extension of the Dirac delta function which we call an *epsilon function* and which has a certain intrinsic interest.

Keywords: Fractal curves, Fractional Integration, Nonstandard Analysis.

1 Introduction

This paper was prepared in response to some speculative work on possible relations between fractal curves and fractional integration which appeared in a paper [1] by N.R.Nigmatullin. The claim made in this paper was that

"A relationship is established between Cantor's fractal set (Cantor's bars)and a fractional integral. The fractal dimension of the Cantor set is equal to the fractional exponent of the integral."

Nigmatullin's paper has been subjected to some critical comment by R.S. Rutman [2,3], who remarks in [3] that

"Is there a relation between fractional calculus and fractal geometry? ..... In the first part of this paper, some recently suggested models are reviewed and no convincing evidence is found for any dynamic model of a fractional order system having been built with the help of fractals. "

Rutman's comments would seem to be justified but the main concern of this paper is to question the feasibility of Nigmatullin's claim from the very outset. The explosion of interest in fractal curves and the ease with which examples can be demonstrated by computer imagery has tended to obscure the fact that these are genuinely complicated objects and that there are real difficulties associated with their mathematical description. Objects which not so long ago would have been described as 'pathological' and treated with due caution are now deceptively familiar. It is salutary to be reminded of the true significance of the terms "function" and "derivative" in connection with singular objects like Cantor sets.

Old familiar problems associated with the Dirac delta function offer a useful parallel, and we make no excuse for re-examining such well explored material. The delta function can be defined as a derivative of the Heaviside unit step function $H(t)$ but only in a generalised sense which distinguishes it from the classical derivative $H'(t)$. Similarly it is possible to present a meaningful description of a generalised derivative of the Cantor "Devil's staircase" function, which has a classical derivative vanishing almost everywhere. To do so we find it convenient to introduce a new mathematical idea, viz. a simple extension of the Dirac delta function which we call an **epsilon function**. This would appear to have some intrinsic interest and value, but it does require some initial acquaintance with the ideas and notation of **Nonstandard Analysis**. Accordingly a brief summary of the relevant material is provided in Appendix 1 To begin with we recall some familiar facts of linear systems theory.

2 Linear systems and delta functions

2.1 Linear systems and convolution

The mathematical representation of a time-invariant linear system (TILS), which carries input signals $x(t)$ into output signals $y(t)$, most generally takes the form of a convolution

$$x(t) \to y(t) = (\mu \star x)(t) \equiv < \mu(\tau), x(t-\tau) > \tag{2.1}$$

where μ is a distribution characterising the system in question. Elementary accounts of systems theory are usually framed in terms of the simpler description which obtains when the characteristic distribution μ is regular, and may be identified with an ordinary (locally integrable) function $g(t)$. In this case (2.1) becomes a classical convolution integral,

$$x(t) \to y(t) = \int_{-\infty}^{+\infty} x(t-\tau)g(\tau)d\tau = \int_{\infty}^{+\infty} x(\tau)g(t-\tau)d\tau. \tag{2.2}$$

The **memory function** $g(t)$ is often referred to as the **impulse response function**, being regarded as the system reponse to a delta function input. Then, as Rutman remarks in [3],

"If $g(t) = \{1 \text{ for } t \geq 0, 0 \text{ for } t < 0\}$ (step function), we have a conventional first-order integration, and if $g(t)$ is the Dirac delta function $\delta(t)$, this transformation amounts to an identical reproduction of the input (the zero-order integration). It is logical to infer that the fractional integration of the order $\nu, 0 < \nu < 1$, will have the memory function interpolating in a sense, between the δ-function and the step function."

That is to say we have on the one hand,

$$x(t) \to y(t) = \int_{-\infty}^{+\infty} x(\tau)H(t-\tau)d\tau = \int_{-\infty}^{t} x(\tau)d\tau \tag{2.3a}$$

and on the other

$$x(t) \to y(t) = \int_{-\infty}^{+\infty} x(\tau)\delta(t-\tau)d\tau = x(t). \tag{2.3b}$$

Strictly speaking, like is not being compared with like here, and there is a real danger of conceptual confusion. While the step function $H(t)$ can legitimately be described as a *function* in the proper sense of the word, this is not the case with the so-called delta function $\delta(t)$ which, in a certain sense, behaves as a 'derivative' of $H(t)$ but which is *not* the classical derivative of that function. The integral in (2.3b) is to be understood in a purely symbolic sense. All this is arguably well known, and problems associated with the definition and use of the delta function are usually assumed to have been entirely cleared up by the introduction of the Schwartz theory of distributions. Nevertheless it is worth closer examination in the present context.

The delta function is still often conceived by the physicist or engineer in terms of the naive description given by Dirac in [4]. That is to say, it is defined as a function $\delta(t)$ satisfying the conditions,

$$\delta(t) = 0 \quad \text{for} \quad t \neq 0 \quad \text{and} \quad \delta(0) = +\infty, \tag{2.4a}$$

$$\int_{-\infty}^{+\infty} x(t)\delta(t)dt = f(0), \tag{2.4b}$$

where in (2.4b) $x(t)$ is any function "which is sufficiently smooth" in the neighbourhood of the origin. However, as is well known, no function exists in standard mathematical analysis which can exhibit both the properties (2.4a) and (2.4b). In the Schwartz theory of distributions the pointwise description (2.4a) is abandoned and the fundamental sampling operation (2.3b) is interpreted by treating δ not as a function, in the proper sense of the word, but as a *functional* acting on a suitable space of test functions. It is not necessary, nor even desirable, to go as far as this in the present context. For systems of the type represented by equation (2.2) there is an alternative formulation in terms of a **Stieltjes convolution integral**:

$$x(t) \to y(t) = \int_{-\infty}^{+\infty} x(t-\tau)dG(\tau) = \int_{-\infty}^{+\infty} x(\tau)dG(t-\tau) \tag{2.5}$$

where

$$G(t) = \int_{-\infty}^{t} g(\tau)d\tau. \tag{2.6}$$

The system is now characterised by the response $G(t)$ to a step function $H(t)$ applied as an input, rather than by the fictitious delta function $\delta(t)$. Moreover, the frequency response which is usually defined as the Fourier transform of the memory function $g(t)$, is equally well specified as the Fourier-Stieltes transform of the step response $G(t)$:

$$\tilde{g}(\omega) = \int_{-\infty}^{+\infty} e^{-i\omega t}g(t)dt \equiv \int_{-\infty}^{+\infty} e^{-i\omega t}dG(t).$$

If $G(t)$ is differentiable then $G'(t) = g(t)$ and is defined everywhere. If $G(t)$ is absolutely continuous then we still have $G'(t) = g(t)$ but the derivative may be defined only almost everywhere. Such a function $G(t)$ defines an absolutely continuous **measure** (or distribution of mass) on the real line $\mathcal{R}$.

More importantly, the Stieltjes integral formulation can still be valid even if no *function* $g(t)$ exists which can act as a memory function. Similarly, the Fourier-Stieltjes transform of $G(t)$ may exist even if there is no function $g(t)$ of which it is the Fourier transform. In this case the function $G(t)$ defines a **singular measure** on $\mathcal{R}$. The most obvious example of this is, of course, the identity operator (or zero-order integrator) for which we have

$$x(t) \to \int_{-\infty}^{+\infty} x(t-\tau)dH(\tau) = \int_{-\infty}^{+\infty} x(\tau)dH(t-\tau) = x(t) \tag{2.7a}$$

and

$$\int_{-\infty}^{+\infty} e^{-i\omega t}dH(t) = 1. \tag{2.7b}$$

The symbol δ can now be legitimately introduced to represent the **Dirac measure** defined on the line by the function $H(t)$. As such it can be conceived as the result of concentrating a unit mass at the origin. The corresponding density function admits the pointwise description attributed to the Dirac delta function in equation (2.4a), but this is purely qualitative and contributes nothing towards the definition of the measure.

If A is any discrete set of points on the line (i.e. any set, finite or infinite, which contains no limit point) then it constitutes a set of Lebesgue measure zero, and a mass distribution concentrated on A can be defined by assigning an appropriately weighted Dirac measure to each point of A. This does not exhaust the possible types of singular measure as the example of the Cantor set makes plain.

2.2 Cantor sets

For the sake of clarity we recall briefly the usual construction of a Cantor subset of the interval $[0,1]$. First, divide the interval $[0,1]$ into three parts of lengths proportional to ξ, $1-2\xi$, and ξ respectively, where $0 < \xi < 1/2$. Remove the central open interval (hereinafter called a "black" interval) of length $1-2\xi$. Then perform the same operation on each of the two remaining "white" intervals, of length ξ. Continuation by iteration of this basic operation generates at the kth step 2^k closed white intervals, each of length ξ^k, and we denote by E_k the set of all points belonging to these intervals. The left-hand end-points of the white intervals may be expressed in the following explicit form:

$$t = j_1(1-\xi) + j_2\xi(1-\xi) + \ldots + j_k\xi^{k-1}(1-\xi), \tag{2.8}$$

where the j_r take only the values 0 or 1. Then the intersection of all the E_k is an uncountably infinite set E whose members are all those points in $[0,1]$ which are given by an infinite series of the form

$$t = \sum_{k=1}^{+\infty} j_k \xi^{k-1}(1-\xi). \tag{2.9}$$

The set E is perfect (i.e. a closed set each of whose points is a limit point) and its measure is

$$\lim_{k\to\infty} (\xi^k 2^k) = 0. \tag{2.10}$$

When ξ is given the particular value $1/3$, E is the well known Cantor ternary set.

Let $p_k(t)$ denote the density function corresponding to a uniform distribution of unit mass over the set E_k. This will be a linear combination of 2^k 'top-hat' functions each of width ξ^k, height $(2\xi)^{-k}$ and area $1/2^k$. Integrating the $p_k(t)$ gives

$$P_k(t) = \int_{-\infty}^{t} p_k(\tau)d\tau$$

which is a continuous function, monotonely increasing from 0 to 1. Its classical derivative $P_k'(t)$ is defined almost everywhere and equal to $p_k(t)$. The pointwise limit

$$P(t) = \lim_{k\to\infty} P_k(t)$$

is a well defined, continuous, monotone increasing function (sometimes called the 'Devil's staircase') which has a classical derivative defined and equal to zero almost everywhere. We can give an explicit definition of $P(t)$ as follows:

For $t = \sum_{k=1}^{+\infty} j_k \xi^{k-1}(1-\xi) \in E$ define $P(t) = \sum_{k=1}^{+\infty} j_k/2^k$.

Then $P(t)$ takes the same value at the end-points of each 'black' interval, and accordingly we define $P(t)$ to have this common value at each interior point. Finally, we set $P(t) = 0$ for $t \leq 0$ and $P(t) = 1$ for $t \geq 1$.

Since, from (2.10), E is a null set any standard function $g(t)$ supported on E, no matter how large its values, will be a null function and every convolution integral

$$\int_{-\infty}^{+\infty} x(t)g(t-\tau)d\tau$$

will vanish identically. On the other hand, since E is uncountable, assigning Dirac delta functions to the points of E, no matter how small their strengths, will always produce an infinite answer. Nevertheless there is a well-defined Stieltjes integral associated with the set E obtained by using the singular function $P(t)$ as integrator, and a corresponding convolution:

$$x(t) \to \int_{-\infty}^{+\infty} x(\tau)dP(t-\tau) \tag{2.11}$$

Moreover $DP(t)$ has a well defined Fourier transform which we can compute as the Fourier-Stieltjes Transform of P(t):

$$\int_{\infty}^{+\infty} e^{-i\omega t} dP(t) = \int_{0}^{1} e^{-i\omega t} dP(t)$$

and we can approximate this by the sum

$$\frac{1}{2^k} \sum \exp\left\{-i\omega\left(j_1(1-\xi) + j_2\xi(1-\xi) + \ldots + j_k\xi^{k-1}(1-\xi)\right)\right\}$$

where the summation extends to the 2^k combuinations of $j_k = 0, 1$. Then

$$\int_{\infty}^{+\infty} e^{-i\omega t} dP(t) = \lim_{k\to\infty} \frac{1}{2^k} \prod_{r=0}^{k-1} (1 + \exp\{-i\omega\xi^r(1-\xi)\})$$

$$= \lim_{k\to\infty} \frac{1}{2^k} \prod_{r=0}^{k-1} e^{-i\omega\xi^r(1-\xi)/2} \left\{e^{i\omega\xi^r(1-\xi)/2} + e^{-i\omega\xi^r(1-\xi)/2}\right\}$$

$$= \lim_{k\to\infty} e^{-i\omega\sum_{r=0}^{k-1}\xi^r(1-\xi)/2} \prod_{r=0}^{k-1} \cos\{\omega\xi^r(1-\xi)/2\}$$

$$= e^{-i\omega/2} \prod_{r=0}^{\infty} \cos\{\omega\xi^r(1-\xi)/2\}$$

since $\sum_{r=0}^{k-1} \xi^r(i-\xi) = 1$.

3 Nonstandard representation

3.1 Pre-delta functions

A relatively straightforward alternative interpretation of singular measures such as those generated by Dirac measures (delta functions) and those concentrated on Cantor-type sets can be offered using the formalism of **Nonstandard Analysis**. A simple ultrapower model of the hyperreal number system $^*R$, together with ad hoc definitions of internal sets and functions is an adequate background for much of the development, and a brief summary of the essential ideas and notation is given in Appendix 1. A fuller account of this form of NSA, together with some of its applications to delta functions, can be found in the final chapter of [5].

Recall first the simplest standard approach to an interpretation of the Dirac symbol $\delta(t)$ as a limit, in a certain sense, of a sequence of suitably chosen ordinary functions. In particular, let $d(t) = 1$ for $|t| < 1/2$ and $d(t) = 0$ for $|t| \geq 1/2$, and for each $k \in \mathcal{N}$ set $d_k(t) \equiv kd(kt)$. Then we may identify the pointwise limit

$$\lim_{k\to\infty} d_k(t) = \begin{cases} +\infty, & \text{for } t = 0 \\ 0, & \text{for } t \neq 0. \end{cases}$$

with the classical derivative of the unit step function $H(t)$. Then for any sufficiently well behaved function $x(t)$ we have

$$\int_{-\infty}^{+\infty} x(\tau)\{\lim_{k\to\infty} d_k(\tau)\}d\tau \equiv \int_{-\infty}^{+\infty} x(\tau)H'(\tau)d\tau = 0, \tag{3.1a}$$

whereas,

$$\lim_{k\to\infty}\int_{-\infty}^{+\infty} x(\tau)d_k(\tau)d\tau = \int_{-\infty}^{+\infty} x(\tau)dH(\tau) = x(0). \tag{3.1b}$$

It is equation (3.1b) which we write symbolically as

$$\int_{-\infty}^{+\infty} x(\tau)\delta(\tau)d\tau = x(0),$$

and the distinction between the ordinary derivative $H'(t)$ of $H(t)$ and the generalised one denoted by $\delta(t)$ is plain.

A typical nonstandard treatment can be summarised briefly as follows: Let $\theta(t)$ be a standard function which is everywhere non-negative and such that

$$\int_{-\infty}^{+\infty} \theta(t)dt = 1, \quad \text{and} \quad \lim_{|t|\to\infty} \theta(t) = 0.$$

For $n = 1, 2, \ldots$ let $\theta_n(t) \equiv n\theta(nt)$ on $\mathcal{R}$, and denote by $\delta_{(\theta)}$ the internal function defined by the equivalence class $[(\theta_n)_{n\in\mathcal{N}}]$. Then $\delta_{(\theta)}$ is a *pre-delta function* in the sense that

$$\delta_{(\theta)}(t) \approx 0, \quad \text{for all} \quad x \notin monad(0), \tag{3.2}$$

and

$$^*\!\int_{-\Lambda}^{+\Lambda} {}^*x(t)\delta_{(\theta)}(x)dx \approx x(0), \tag{3.3}$$

for every continuous standard function $x(t)$ and every positive infinite hyperreal Λ.

In particular, taking for the kernel function θ the simple rectangular pulse $d(t) = 1$ for $|t| < 1/2$ and $d(t) = 0$ otherwise, we get a nonstandard model for $\delta(t)$ which comes closest to the intuitive picture suggested by Dirac's original description:

$$\delta_{(d)}(t) = \begin{cases} 0 \ , & \text{for} \ \ t \geq 1/2\Omega \\ \Omega \ , & \text{for} \ \ |t| < 1/2\Omega \\ 0 \ , & \text{for} \ \ t \leq -1/2\Omega, \end{cases}$$

where, as usual, we use the special symbol Ω for the infinite hyperreal $[n]$.

The symbol δ now admits an interpretation in the standard mathematical universe as an operator (functional) obtained by taking the standard part of a genuine integration process in $^*\mathcal{R}$:

$$< \delta, x >= \int_{-\infty}^{+\infty} x(t)\delta(t)dt \equiv st\left\{{}^*\!\int_{-\Lambda}^{+\Lambda} {}^*x(t)\delta_{(\theta)}(t)dt\right\} \tag{3.4}$$

where $\delta_{(\theta)}$ is any pre-delta function and Λ is any positive infinite hyperreal. There is similarly no difficulty in defining a delta function of arbitrary strength k, where k is any given number in $\mathcal{R}$. We have only to choose a kernel function θ which is such that $\int_{-\infty}^{+\infty} \theta(t)dt = k$, and then the internal function defined by the sequence $(\theta_n)_{n\in\mathcal{N}}$ generates a sampling operation which carries x into $kx(0)$.

3.2 Epsilon functions

Now suppose we try to compute a corresponding 'density function' for the mass distribution associated with the Cantor staircase function $P(t)$. We have,

$$p_k(t) = \sum_{y\in E_k} \frac{1}{2^k} d(kt - y), \tag{3.5}$$

which gives a pointwise limit of the form

$$\lim_{k\to\infty} = \begin{cases} +\infty, & \text{for } t \in E, \\ 0, & \text{otherwise} \end{cases}$$

and this we may identify with the classical derivative of the Cantor staircase function $P(t)$. Moreover we have

$$\int_{-\infty}^{+\infty} x(\tau)\{\lim_{k\to\infty} p_k(\tau)\}d\tau \equiv \int_{-\infty}^{+\infty} x(\tau)P'(\tau)d(\tau) = 0$$

whereas

$$\lim_{k\to\infty} \int_{-\infty}^{+\infty} x(\tau)p_k(\tau)d\tau = \int_{-\infty}^{+\infty} x(\tau)P(\tau),$$

which we would like to write symbolically as a generalised derivative of $P(t)$ in the same way as $\delta(t)$ is the generalised derivative of $H(t)$. That is, we would like to be able to write

$$DP(t) = \int_{-\infty}^{+\infty} x(\tau) \left\{ \sum_{y\in E} \frac{1}{2^\Omega} \delta(t-\tau) \right\} d\tau, \tag{3.6}$$

where Ω is infinite, so that each delta function has a (specific) infinitesimal weight, and the summation extends over an *uncountable* infinity of points.

There is no way in which we can formulate this description of a generalised derivative of $P(t)$ in standard mathematical terms. However, within the nonstandard context it makes sense to deal with internal functions of the form $\epsilon\delta_{(\theta)}(t)$ where $\epsilon \approx 0$. Such a pre-delta function of infinitesimal strength appears, at first, to have no standard interpretation. A sampled value $\epsilon x(0)$ is indistinguishable from zero in the standard universe, and a nonstandard equation of the form

$$Dx(t) = x'(t) + \epsilon\delta_{(\theta)}(t-a),$$

reduces to the identity $Dx(t) \equiv x'(t)$ when we restrict t to $\mathcal{R}$ and take standard parts. Nevertheless there is some value in considering such internal functions and investigating their effect on standard functions. For convenience we shall use the general notation $\epsilon(t)$ for a pre-delta function of (arbitrary) infinitesimal strength and refer to it as an **epsilon function.**

The properties of the Cantor staircase function $P(t)$ are made more readily comprehensible by an alternative approach set in the context of the hyperreal line $^*\mathcal{R}$. Consider again the set E_k defined at the kth step of the construction of E. The density function $p_k(t)$ defined in (3.5) is a train of 2^k "top-hat" functions, each of area $1/2^k$. Now denote by Ψ the internal function defined by the sequence $(p_k)_{k\in\mathcal{N}}$. It is readily seen that $\Psi(t)$ is a hyperfinite sum of 2^Ω epsilon functions, each of (infinitesimal) strength $1/2^\Omega$,

$$\Psi(t) = \sum_{y\in[E_k]} \frac{1}{2^\Omega}\delta_{(d)}(t-y) \tag{3.7}$$

The shadow of $\Psi(t)$ in $\mathcal{R}$ is identically zero. Nevertheless if we integrate $\Psi(t)$ and take the standard part we obtain a function $\Phi(t)$ whose shadow on $\mathcal{R}$ does *not* vanish identically. In fact we have

$$^*\int_{-\Omega}^{t} \Psi(\tau)d\tau = \Phi(t) \approx P(t)$$

on the finite part of $^*\mathcal{R}$.

The singular behaviour of $P(t)$ lies in the fact that although it has a classical derivative which vanishes almost everywhere and has no (finite) jumps, it nevertheless manages to increase from zero to 1 over the real interval $[0,1]$ - behaviour which seems particularly disconcerting if we try to interpret its graph as a space/time plot. From the nonstandard viewpoint, however, $P(t)$ has a $*$-derivative consisting of infinitely many epsilon functions and therefore $P(t)$ is able to increase by virtue of infinitely many infinitesimally small jumps, not manifest in the standard picture.

3.3 Extreme values for ξ

The parameter ξ has been restricted to values in the open real interval, $(0, 1/2)$, as is usual. If we allow ξ to assume hyperreal values then it becomes possible to discuss the limiting cases as ξ approaches 0 or $1/2$. For $\xi \approx 0$ the epsilon functions comprising $\Psi(t)$ are all concentrated in the monads of 0 and 1, and this is effectively equivalent to having a delta function of strength $1/2$ at 0 and at 1. The corresponding standard function $P(t)$ is no longer a continuous singular function; it has the constant value $1/2$ for $0 < t < 1$ and has simple jump discontinuities at $t = 0$ and $t = 1$. For $\xi \approx 1/2$ the epsilon functions of $\Psi(t)$ are uniformly distributed throughout $^*[0/1]$ at points of the form $p/2^k$. The standard function $P(t) = t$ throughout the real interval $[0,1]$ and is a straightforward continuous function which is differentiable everywhere in the classical sense except at the points 0 and 1 themselves.

This is in general agreement with the discussion given by Nigmatullin in section (3) of [1]. The singular behaviour of $P(t)$ depends on the presence of epsilon functions in its (generalised) derivative, and on the way in which those epsilon functions are distributed. However there is nothing which suggests any correlation with fractional integration or fractional differentiation, as Nigmatullin purports to establish. Nigmatullin's analysis of his "fractional integral' is effectively equivalent to our construction of the internal function $\Psi(t)$, but he is unable to proceed with it because of the constraints imposed by standard mathematical analysis. In effect he does not have the formal apparatus to discuss this adequately. Hence he takes instead the Laplace transform of the approximating functions $p_k(t)$ and then proceeds to the limit to obtain a certain recurrence relation,

$$\tilde{Q}(z/\xi) = 1/2\tilde{Q}(z).$$

However, it does not follow that the solution of this functional equation must be of the form

$$\tilde{Q}(z) = A_\nu z^{-\nu}$$

as he asserts. It is true that *if* we consider a solution of this form then we necessarily have $\nu = \ln 2/\ln(1/\xi)$, but this simply begs the question. Finally, to achieve a characteristic resembling the desired power law he brings in a multiplicative factor $1/t$, which is inconsistent with the convolution form of the fractional integration operator which it is required to achieve.

Appendix 1

Nonstandard Analysis appeared in the 1960s as a consequence of the work of the mathematical logician Abraham Robinson, and his text [5] is still the basic authoritative source for the subject. Other, more readily accessible expositions have since appeared and the recent book [6] by Robert Goldblatt is particularly recommended. The following account summarises very briefly the essential ideas and notations used in the main text of this paper.

Standard mathematical analysis deals with a universe of objects each of which is ultimately definable in terms of the set R of the real numbers. Thus, for example, a real function f mapping a set A into $\mathcal{R}$, is defined by, and may be identified with, the set

$$A_f = \{(x, y) : x \in A \quad \text{and} \quad y = f(x)\}$$

where each ordered pair (x, y) can itself be defined as a certain set $\{\{x\}, \{x, y\}\}$ whose elements are themselves real numbers. Similarly, an integral such as $\int_0^1 f(x)dx$ can be defined as a certain *functional*, (that is, a mapping of functions into numbers) and therefore again be identified with a set of ordered pairs, this time of functions and numbers. Since each function can be identified with a specific set of ordered pairs of reals, and therefore defined in terms of the real numbers, so also can the integral itself. And similarly any other object of standard mathematical analysis can be formally defined as a superset, of some finite order, based on $\mathcal{R}$. Call the whole collection $V(\mathcal{R})$ of all such sets *the standard real universe.*

Nonstandard Analysis deals with a universe $V({}^*\mathcal{R})$ which is a like collection of sets based on an enlarged version ${}^*\mathcal{R}$ of the real number system. This contains infinite and infinitesimal elements as well as (copies of) the real numbers themselves. The members of ${}^*\mathcal{R}$ are called *hyperreal numbers* and, in a certain sense, they inherit all the properties of the real numbers. The enlargement of R to ${}^*\mathcal{R}$ is required to be such that to each 'standard' object α belonging to $V(\mathcal{R})$ there corresponds an object ${}^*\alpha$ in the nonstandard universe $V({}^*\mathcal{R})$ which is of *the same type as* α; ${}^*\alpha$ is called *the nonstandard extension of* α. Thus to each set A of real numbers there corresponds a set ${}^*A$ of hyperreal numbers; in general ${}^*A$ will be a proper extension of A. In the same way, to each function f mapping A into $\mathcal{R}$ there corresponds a function ${}^*f$ mapping ${}^*A$ into ${}^*\mathcal{R}$; the restriction of ${}^*f$ to $\mathcal{R}$ coincides with f. And similarly for all other standard objects in $V(\mathcal{R})$.

The key assumption in NSA is that the enlargement of $\mathcal{R}$ is rich enough to ensure that there is a *Transfer Principle* relating the standard universe $V(\mathcal{R})$ and the nonstandard universe $V({}^*\mathcal{R})$. This says that any well-formed statement (theorem) about standard objects $\alpha, \beta, \ldots$, in $V(\mathcal{R})$ will be true if and only if the corresponding statement about their nonstandard images ${}^*\alpha, {}^*\beta, \ldots$ is true in $V({}^*\mathcal{R})$. It follows that theorems about standard objects in $V(\mathcal{R})$ can often be established quickly and easily by working in terms of their nonstandard extensions and exploiting the richer structure of $V({}^*\mathcal{R})$. Moreover there exist so-called 'internal functions', defined on and taking values in ${}^*\mathcal{R}$, which need not be the nonstandard extensions of any ordinary functions but *to which, using the Transfer Principle, we can rigorously apply versions of the familiar processes and theorems of the calculus.* In particular there are internal functions which can be interpreted as representatives of various types of delta function.

A brief outline of a simple form of such a nonstandard universe has been given elsewhere [7]. In this a model for the hyperreal number system ${}^*\mathcal{R}$ is defined in terms of equivalence classes $[(x_n)_{n\in\mathcal{N}}] \equiv [x_n]$ of infinite sequences of real numbers $(x_n)_{n\in\mathcal{N}}$. Internal subsets Γ of ${}^*\mathcal{R}$ are then defined as equivalence classes of sequences $(G_n)_{n\in\mathcal{N}}$ of standard subsets of $\mathcal{R}$, and internal functions F as equivalence classes of sequences $(f_n)_{n\in\mathcal{N}}$ of standard functions on $\mathcal{R}$. In particular the standard set $\mathcal{N}$ of the positive integers admits a nonstandard extension to a set ${}^*\mathcal{N}$ which contains infinite integers such as $\Omega = [n]$, $\Omega^2 = [n^2]$ and in general $\Lambda = [(m_n)_{n\in\mathcal{N}}]$, where the m_n are standard positive integers. This allows the extension of a standard sequence such as $(a_n)_{n\in\mathcal{N}}$ to the nonstandard $(a_\nu)_{\nu\in{}^*\mathcal{N}}$. Taking a sequence of (standard) partial sums $\left(\sum\limits_{k=-n}^{n} a_k\right)_{n\in\mathcal{N}}$ for example, we obtain so-called *hyperfinite* sums of the form

$$\sum_{k=-\Lambda}^{\Lambda} a_k = \left[\left(\sum_{k=-m_n}^{m_n} a_k \right)_{n \in \mathcal{N}} \right].$$

Such hyperfinite sums enjoy all the algebraic properties appropriate to finite sums.

Further, in order to develop a calculus for internal functions we can define a nonstandard *-differential operator $^*D$ and a corresponding nonstandard elementary *-integral as follows:

$$^*DF = [(Df_n)_{n \in \mathcal{N}}], \tag{A.1}$$

and

$$^*\int_\alpha^\beta F(x)dx = \left[\left(\int_{a_n}^{b_n} f_n(x)dx \right)_{n \in \mathcal{N}} \right]. \tag{A.2}$$

Then standard theorems from elementary calculus (such as mean value theorems, partial integration formulae, etc.) transfer to these concepts in a fairly obvious way.

In the main text we have occasion to use the symbol $\approx$ which denotes the relation 'infinitely close' between hyperreal numbers: we write $x \approx y$ whenever $x - y$ is infinitesimal. The *standard part* of a finite hyperreal number x is the (unique) real number $r = st(x)$ which is such that $x \approx r$. The set of all hyperreal x infinitely close to a given real number r will be called the *monad* of r, and we write

$$\text{mon}(r) = \{x \in {}^*R : x \approx r\}.$$

A key feature of NSA is the way in which the operation of taking the limit reduces to that of taking the standard part. Thus, for example,

$$\lim_{x \to a} f(x) = f(a) \quad \text{if and only if} \quad f(a) = st\{{}^*f(x)\} \quad \text{for all} \quad x \approx a.$$

Similarly the infinite series $\sum_{k=\infty}^{+\infty} a_k$ converges to the real number a if and only if for every positive infinite integer Λ we have

$$st\left\{ \sum_{k=-\Lambda}^{+\Lambda} a_k \right\} = a.$$

Bibliography

1. N.R. Nigmatullin Fractional Integral and its Physical Interpretation. *Theor. and Math. Phys., vol.90, pt.3, 242-251 (1992).*

2. R.S. Rutman *Teor. Mat. Fiz., 100, 476-478 (1994)*

3. R.S. Rutman On Physical Interpretations of Fractional Integration and Differentiation. *Theor. and Math. Phys., 105, No.3, 1509-1519 (1995).*

4. P.A.M. Dirac *The Principles of Quantum Mechanics*, C.U.P. (1930).

5. A. Robinson *Nonstandard Analysis*, North Holland (1966).

6. R. Goldblatt *Lectures on the Hyperreals*, Springer (1998).

7. R.F. Hoskins *Delta Functions*, Horwood Publishing Ltd., Chichester (1999).

Diffusion on Fractals – Efficient algorithms to compute the random walk dimension

A. Franz[1], **C. Schulzky**[2], **S. Seeger**[1], **K.H. Hoffmann**[1]

[1]Institut für Physik, Technische Universität, D-09107 Chemnitz, Germany
[2]Department of Applied Mathematics, The University of Western Ontario, London N6A 5B, Canada

Abstract

Self-similar fractals are used as a simple model for porous media in order to describe diffusive processes. The diffusion or Brownian motion of particles on a fractal is approximated by random walks on pre-fractals. Since there are a lot of holes in the fractal, where a random walker is not allowed to move in, the mean square displacement scales with time t asymptotically as t^{2/d_w}, where the random walk dimension d_w is usually greater than 2. This dimension is an important quantity to characterize diffusion properties.

In this paper three efficient methods to calculate the random walk dimension of finitely ramified Sierpinski carpets are presented: First a simulation of random walks on pre-carpets, where an efficient storing scheme decreases the needed amount of memory and speeds up the computation. Secondly we iterate the master equation describing the time evolution of the probability distribution. Thirdly a resistance scaling algorithm is presented which yields a resistance scaling exponent. This exponent is related to the random walk dimension via the Einstein relation, using analogies between random walks on graphs and resistor networks.

1 Introduction

Diffusion processes in disordered media such as porous materials are widely studied in the physical literature [1, 2]. These processes exhibit an anomalous behaviour in terms of the asymptotic time scaling of the mean square displacement of the diffusive particles:

$$\langle r^2(t)\rangle \sim t^{\gamma}, \tag{1.1}$$

where $r(t)$ is the distance of the particle from its origin after time t. In porous media the diffusion exponent γ is less than one, describing a slowed down diffusion compared to the linear time behaviour known for normal diffusion.

To simulate these kinds of diffusion one has to take the partially self-similar structure of the substrate into account, i.e. holes on a large range of length scales.

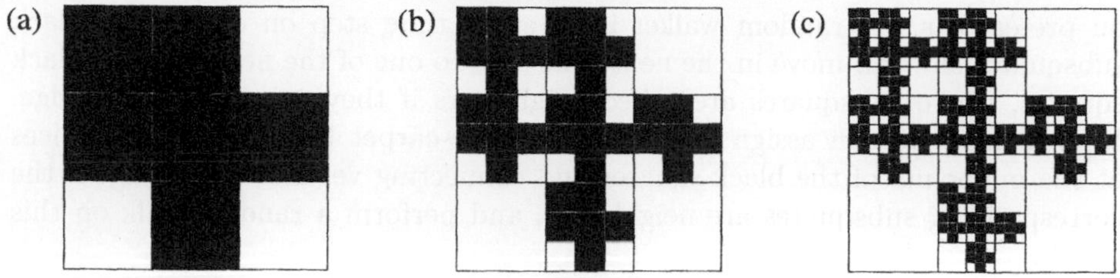

Figure 1. Example of a Sierpinski carpet generator (a) and the results of the second (b) and third (c) construction step of the resulting carpet.

This can be done by modelling the disordered medium as a percolation cluster [2] or as a more regular fractal, like e.g. the Sierpinski carpet family.

Sierpinski carpets are determined by a generator. A *generator* is a square, divided into $n \times n$ congruent subsquares, and m of the subsquares are black and the other $n^2 - m$ are white. Figure 1a shows an example of a generator for a Sierpinski carpet with $n = 3$ and $m = 6$. The construction procedure described by the generator is as follows: Starting with a square in the plane, divide it into $n \times n$ smaller congruent squares and erase all the squares which are marked white in the generator. Thus the black subsquares remain. In the next step every one of the remaining smaller squares is again divided into $n \times n$ smaller subsquares and the ones corresponding to white squares in the generator are erased. This construction procedure can be continued ad infinitum, and the limit object is a fractal called *Sierpinski carpet* with a fractal dimension $d_f = \ln(m)/\ln(n)$ [3]. Figure 1b and c shows the construction steps two and three of the Sierpinski carpet described by the generator shown in figure 1a. The result of finitely many construction steps we call a *pre-carpet.*

Sierpinski carpets may be finitely ramified or infinitely ramified. A carpet is called *finitely ramified* if any subset can be separated from the rest by cutting a finite number of connections, otherwise it is called *infinitely ramified.* Since a Sierpinski carpet is fully described by the generator, the ramification can be directly seen there: In the generator of a finitely ramified carpet the first and the last row coincide in exactly one black subsquare, and the same holds true for the first and last column. If the first and last row or the first and last column of the generator coincide in more than one black subsquare, then the corresponding carpet is infinitely ramified.

In the following we restrict ourselves to finitely ramified Sierpinski carpets. For these carpets we call a subsquare in the first column of the generator *critical* to the left, if this is the black subsquare in which this column coincides with the last column. Analogously critical squares in the last column and the first and last row, i.e. in the directions right, up and down, are defined. By definition a finitely ramified carpet generator has exactly one critical square in each direction. Two of these squares may coincide, so there may be three or four critical squares.

The diffusion process on Sierpinski carpets can be modelled by random walks on pre-carpets. A random walker is at every time step on one of the black subsquares and can move in the next time step to one of the neighbouring black squares, where subsquares are called neighbours if they coincide in one edge. Equivalently, we can assign a graph to the pre-carpet by putting the vertices at the midpoints of the black squares and connecting vertices by an edge if the corresponding subsquares are neighbours, and perform a random walk on this graph.

The random walk dimension is defined via equation (1.1) as

$$d_w = 2/\gamma. \tag{1.2}$$

This dimension is an important quantity to characterize diffusion properties. On an Euclidean lattice we have $d_w = 2$ independent of the dimension of the Euclidean space. But on fractals normally $d_w > 2$ corresponding to a slowed down diffusion.

The simplest way to determine d_w is by simulating a lot of random walks to get the mean square displacement as a function of time. In a log-log-plot the slope of a fitted straight line gives the diffusion exponent γ and hence via equation (1.2) the random walk dimension. An efficient algorithm for this simulation is given in section 2. Another possibility is to iterate the master equation describing the random walk in order to get the probability distribution of the walkers as a function of time, as will be described in section 3. From this distribution the mean square displacement can be extracted. A third possibility, explained in section 4, is to use analogies between random walks on graphs and resistor networks in form of the Einstein relation connecting a resistance scaling exponent and the random walk dimension.

2 Effective simulation of random walks

When performing random walks on pre-fractal graphs normally long times have to be considered until a linear behaviour in the log-log-plot of the mean square displacement over time can be observed. Long walks require large graphs. If the sufficiently large pre-carpets have to be stored in the computer, then the amount of available memory restricts the size of the pre-carpet and thus the number of possible steps until a walker may reach the boundary. If longer walks are considered, then all the walkers reaching the boundary have to be discarded in the averaging process. This will result in undesirable 'surface effects'. The memory problem is addressed by Dasgupta et al. [4]. There an algorithm is presented which stores only the generator but requires a lot of computations to decide for a given point whether it belongs to the fractal or not.

Our algorithm [5] also takes only the generator as input, but reduces the necessary computations per step by employing a hierarchical coordinate notation and pre-computed tables to determine neighbourhood relations. Let us consider the i-th pre-carpet as a combination of m copies of the $(i-1)$-th carpet, ..., and the second pre-carpet as a combination of m copies of the first pre-carpet,

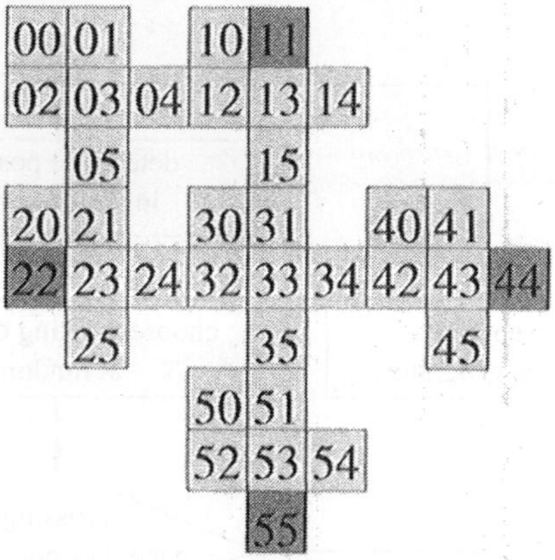

Figure 2. The position inside the i-th pre-carpet is described by an i-dimensional position vector. The dark squares are critical, there a walker might cross the pre-carpet boundaries.

where the first pre-carpet coincides with the generator. To give the position of the walker within the generator, we enumerate the black squares as $p^1 \in \{0, 1, \ldots, m-1\}$. The position $p^i \in \{0, 1, \ldots, m-1\}^i$ within the i-th pre-carpet can then be given as an i-tuple (the position vector) $p^i = (p_1, \cdots, p_i)$, as is illustrated in figure 2 for the example carpet shown in figure 1. In the subsequent text we will refer to the various p_j, $j \in \{1, \ldots, i\}$ as positions in the *j-th level* of the carpet. Within this base-m notation, moves within a generator change only p_1, the position at the lowest level of the carpet.

The algorithm for one walker step is given in figure 3. We first determine the current walker position. For the first step we choose at random a number $p_1 \in \{0, \ldots, m-1\}$, starting with a position vector of length 1 that will grow larger as needed.

Then we choose a direction to walk in at random and check if that might cause the walker to cross the generator boundaries. There may be 1 to 4 possible directions, depending on the position within the generator. We therefore use a 12 slot table with the possible directions repeated according to their probability. For the example carpet shown in figure 1 we will obtain table 1. Moves within the generator are denoted by normal arrows ($\uparrow$, $\rightarrow$, $\downarrow$, $\leftarrow$), while double arrows ($\Uparrow$, $\Rightarrow$, $\Downarrow$, $\Leftarrow$) denote moves across a boundary.

If we move inside the generator (i.e. we got a normal arrow as direction), we accept the choosen direction, updating the position p_1 accordingly. This is done using another table that – for each allowed position within the generator – gives the position of the neighbour in each direction, assuming a periodic lattice of generators (table 2). The sign $\times$ denotes no neighbour in that particular direction. The sign ! can be neglected here, it is needed later.

If we chose a direction crossing the generator boundaries (i.e. we got a double

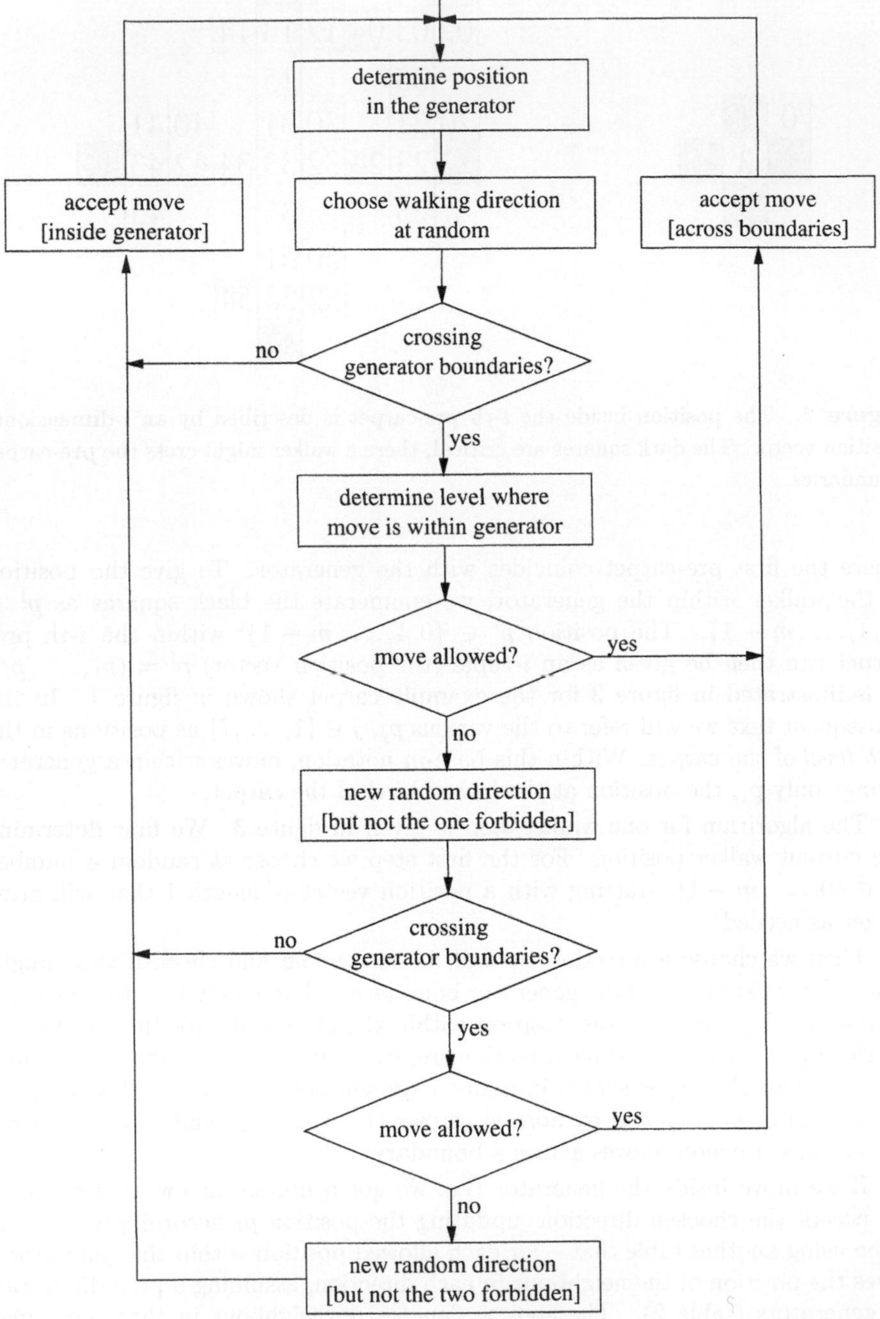

Figure 3. Flowchart to execute one walker move.

0 :	→	→	→	→	→	→	↓	↓	↓	↓	↓	↓
1 :	⇑	⇑	⇑	⇑	↓	↓	↓	↓	←	←	←	←
2 :	↑	↑	↑	↑	→	→	→	→	⇐	⇐	⇐	⇐
3 :	↑	↑	↑	→	→	→	↓	↓	↓	←	←	←
4 :	⇒	⇒	⇒	⇒	⇒	⇒	←	←	←	←	←	←
5 :	↑	↑	↑	↑	↑	↑	⇓	⇓	⇓	⇓	⇓	⇓

Table 1. Direction map for the generator in figure 1.

	0	1	2	3	4	5
↑	×	5!	0	1	×	3
→	1	×	3	4	2!	×
↓	2	3	×	5	×	1!
←	×	0	4!	2	3	×

Table 2. Neighbour map for the generator in figure 1.

arrow), we check whether the neighbouring generator in the chosen direction belongs to the fractal. This can be decided in the j-th level with the property $(p_j \neq p_1) \wedge (p_l = p_1 \ \forall l < j)$. If these positions up to p_j have not yet been determined, we append random numbers out of $\{0, \ldots, m-1\}$ to the position (and origin) vector, thereby refining the position of the walker in the fractal. In this way the accessible carpet pattern becomes effectively infinite, since always when reaching the boundary of the known region we append a number to the position vector describing the position in the next level.

If the position within the j-th level determines that there is no neighbour in the given direction, we have to choose an alternative under the restriction, that the discarded direction is not possible. We use again a 12 slot table, which for the example carpet is shown in table 3. For the special case of three critical squares in the generator this might again result in a direction that may be forbidden due to a non-existent neighbour, in this case we have again to choose

$\not\uparrow$:	↓	↓	↓	↓	↓	↓	←	←	←	←	←	←
$\not\rightarrow$:	←	←	←	←	←	←	←	←	←	←	←	←
$\not\downarrow$:	↑	↑	↑	↑	↑	↑	↑	↑	↑	↑	↑	↑
$\not\leftarrow$:	↑	↑	↑	↑	↑	↑	→	→	→	→	→	→

Table 3. Alternative directions for the generator in figure 1.

an alternative. After at most two times choosing an alternative finally we get an allowed direction. We accept the move and update the position vector.

Moves inside the generator can thus be performed very quickly, whereas the critical squares require some overhead. The fraction of critical squares can be made smaller by using an a few times iterated generator instead of the generator itself, as can be seen in figure 2.

Using the look-up tables and the hierarchical coordinate notation described above we can implement the algorithm very effectively. The resulting program benefits from high data locality, a very small inner loop and few 'unexpected' jumps. A performance study of the algorithm is given in [5].

3 Master equation approach

The random walk investigated in the previous section can be characterized by a master equation

$$P(t+1) = \Gamma \cdot P(t), \tag{3.1}$$

where $P(t)$ denotes the probability distribution that a walker is at a certain position at time t and Γ is the transition matrix. Thus the random walk is a stationary discrete Markov process. Compared to performing random walks iterating the master equation (3.1) has the advantage that a statistical average over a lot of walkers is not necessary any more. But of course now memory is needed for every position covered by a positive probability. In order to keep the needed memory as small as possible in our algorithm [6] we store the fractal dynamically.

The smallest storing unit is a structure containing all information for one generator. For every black square a new and an old probability $P_{i,\mathrm{new}}$ and $P_{i,\mathrm{old}}$ are stored. Arrays a and b contain for every direction (right, left, up, down) information about the transition probabilities to and from the critical squares in this direction. Furthermore an array q of four pointers to the probability values of the corresponding critical squares in the neighbouring generators is stored. A vector p describes the position of the generator in the whole fractal, in a similar way than used in the random walk algorithm. The transition probabilities from non-critical squares are stored in a global array c outside the generator data structure.

All needed generator structures are put together into a linked list. At the beginning this list contains just one generator in which the starting square is situated. This square gets the probability 1, whereas the probabilities of all the other squares are set to zero. The list of generators will be enlarged if squares outside the ones for which memory is allocated get positive probability.

In order to update the probability value of a critical square it has to be known whether the corresponding outside neighbour exists. If the neighbour exists then its probability is needed. This information is stored in the q array of pointers. If such a pointer is not set, then the neighbouring generator in the

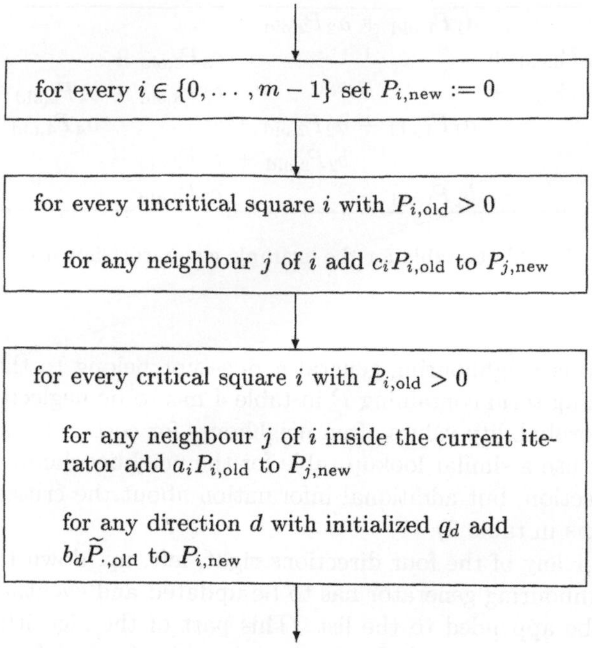

Figure 4. Flowchart to execute the update of the probabilities in one time step for one generator in the list.

considered direction does not exist in the list. There are two reasons for this case: The neighbouring generator does not belong to the fractal, or it belongs to the fractal, but the memory for this generator has not yet been allocated. To distinguish between these two cases the b array is used, as described below.

The a array contains the transition probability for leaving the critical squares of the considered generator, i.e. they depend on the number of neighbours of the critical squares and thus on the existence of the neighbouring generator. One entry in the b array is the transition probability to the corresponding critical square of the considered generator, i.e. from the neighbouring critical square in the neighbouring generator, if this generator belongs to the fractal and belongs already to the list. Otherwise the element in the b array gets the value 0 or -1. Zero means that the neighbouring generator does not belong to the fractal. The value -1 indicates, that the neighbouring generator belongs to the fractal, but has not yet been situated in the list.

In every time step the list of generators is processed from begin to end. The core algorithm for every list element consists of two parts. First the probability values have to be updated, as is shown in a flowchart (figure 4). The resulting linear equation system for the example carpet shown in figure 1 is given in table 4. Note that $\widetilde{P}$ refers to the probability value of the corresponding neighbouring

$P_{0,\text{new}} :=$		$a_1 P_{1,\text{old}} +$	$a_2 P_{2,\text{old}}$			
$P_{1,\text{new}} :=$	$c_0 P_{0,\text{old}} +$			$c_3 P_{3,\text{old}} +$		$b_5 \widetilde{P}_{5,\text{old}}$
$P_{2,\text{new}} :=$	$c_0 P_{0,\text{old}} +$			$c_3 P_{3,\text{old}} +$	$b_4 \widetilde{P}_{4,\text{old}}$	
$P_{3,\text{new}} :=$		$a_1 P_{1,\text{old}} +$	$a_2 P_{2,\text{old}} +$		$a_4 P_{4,\text{old}} +$	$a_5 P_{5,\text{old}}$
$P_{4,\text{new}} :=$			$b_2 \widetilde{P}_{2,\text{old}} +$	$c_3 P_{3,\text{old}}$		
$P_{5,\text{new}} :=$		$b_1 \widetilde{P}_{1,\text{old}} +$		$c_3 P_{3,\text{old}}$		

Table 4. Neighbour table for the example generator shown in figure 1.

generator. If this neighbouring generator does not belong to the fractal, then the corresponding term containing $\widetilde{P}$ in table 4 has to be neglected. During the update of the probability values often neighbours for certain squares have to be computed. We use a similar lookup table for the neighbourhood structure as in the previous section, but additional information about the critical directions is included (! signs in table 2).

Secondly for any of the four directions right, left, up, down the information about the neighbouring generator has to be updated and eventually a new generator has to be appended to the list. This part of the algorithm is shown in figure 5. As in figure 4 d stands for the direction ($d \in$ {right,left,up,down}). The acronym d_{op} means the opposite direction to d, and i denotes the number of the critical square in direction d. All symbols with a tilde refer to neighbouring generators.

As explained before, the test $b_d = -1$ is true if the neighbouring generator in direction d belongs to the fractal, but has not yet been appended to the list. In the case $P_{i,\text{new}} > 0$ during the next time step probability will flow from the current generator to this neighbour. Thus this neighbouring generator has to be initialized.

To initialize a neighbouring generator its position vector $\widetilde{p}$ has to be determined. The neighbour table gives for p_1 and the considered direction a value $\widetilde{p}_1$ and maybe a ! sign. If we get just a $\widetilde{p}_1$ without !, then the position vector of the neighbouring generator coincides with the array of the current generator except in p_1, i.e. $\widetilde{p}_i = p_i$ for any $i > 1$. If we get a ! sign then $\widetilde{p}_2$ differs from p_2. Thus we have to determine $\widetilde{p}_2$ as well by looking in the neighbour table. As long as we get a ! sign we have to continue to look for the next position entry.

In a Sierpinski carpet with loops it may be possible that probability flows from different directions in a generator. To avoid that in such a case the generator is appended several times to the list, before appending it is searched for the generator with the given position vector. If the generator is found then the arrays a, b and q have to be updated. Otherwise additionally memory for the generator with the given position vector has to be allocated, and this generator has to be appended to the list.

This algorithm for one time step can be repeated until no memory is left on

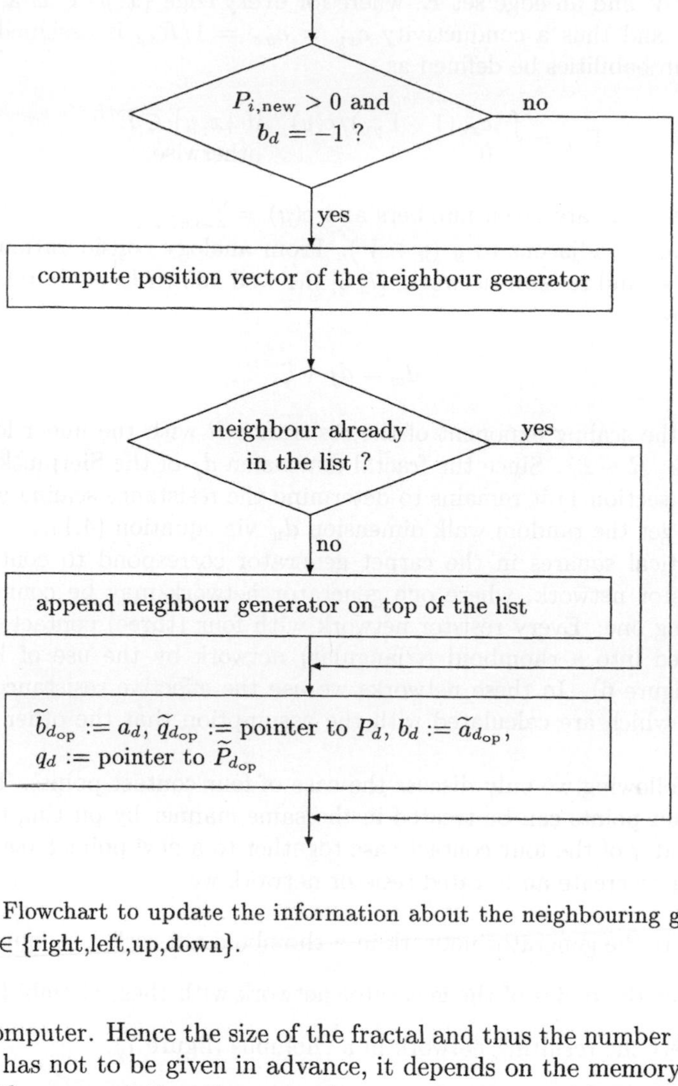

Figure 5. Flowchart to update the information about the neighbouring generator in direction $d \in$ {right,left,up,down}.

the used computer. Hence the size of the fractal and thus the number of possible time steps has not to be given in advance, it depends on the memory available. Of course the computation time per time step increases with increasing time, since the list of generators which has to be processed becomes longer and longer.

4 Resistance scaling method

Diffusion processes and the current flow through an adequate resistor network are strongly connected [7]. Assigning a unit resistance to every edge of the graph representation of a Sierpinski pre-carpet (see section 1) we get the corresponding resistor network. On the other hand we can assign a random walk to a given resistor network, described by an undirected connected graph $G = (V, E)$ with a vertex set V and an edge set E, where for every edge $\{x, y\} \in E$ a resistance $R_{xy} = R_{yx}$ and thus a conductivity $c_{xy} = c_{yx} = 1/R_{xy}$ is assigned: Let the transition probabilities be defined as

$$\Gamma_{xy} = \begin{cases} c_{xy}(1-\Gamma_{yy})/c(y) & \text{if } \{x,y\} \in E, \\ 0 & \text{otherwise,} \end{cases}$$

where $\Gamma_{yy} \in [0, 1)$ are given numbers and $c(y) = \sum_{w:\{w,y\}\in E} c_{wy}$ is the sum of all conductances adjacent to y ($y \in V$). From analogy considerations between random walks and resistor networks [7, 8] the well known Einstein relation [2, 9, 10, 11] follows:

$$d_w = d_f + \zeta, \tag{4.1}$$

where ζ is the scaling exponent of the resistance R with the linear length L of the network: $R \sim L^{\zeta}$. Since the fractal dimension d_f of the Sierpinski carpet is known (see section 1) it remains to determine the resistance scaling exponent ζ in order to get the random walk dimension d_w via equation (4.1).

The critical squares in the carpet generator correspond to contact points in the resistor network, where one generator network may be connected to a neighbouring one. Every resistor network with four (three) contact points can be converted into a rhomboid (triangular) network by the use of Kirchhoff's laws (see figure 6). In these networks we use the effective resistances between two points which are calculated with the assumption that the other points are isolated.

In the following we only discuss the case of four contact points. The case of three contact points can be treated in the same manner by putting the contact points β and γ of the four contact case together to a new point ξ (see figure 6).

In order to create an iterated resistor network we

1. convert the generator network in a rhomboid network (see figure 6c),
2. replace the nodes of the generator network with these rhombi (figure 7),
3. convert the resulting network in a rhombus (figure 7),
4. repeat steps 2 and 3.

This procedure defines a function $\mathbf{R}' = f(\mathbf{R})$, which is independent of the iteration step. $\mathbf{R}$ and $\mathbf{R}'$ denote the resistance vector for the rhombus before and after one iteration step. The components of $\mathbf{R}$ are given by

$$\mathbf{R} = (R_{\alpha\beta}, R_{\alpha\gamma}, R_{\alpha\delta}, R_{\beta\gamma}, R_{\beta\delta}, R_{\gamma\delta}).$$

The in general different ratios $\lambda_i = f_i(\mathbf{R})/R_i$, $i \in \{\alpha\beta, \alpha\gamma, \alpha\delta, \beta\gamma, \beta\delta, \gamma\delta\}$ converge to one unique value, λ, for suffiently large iterations. This means that

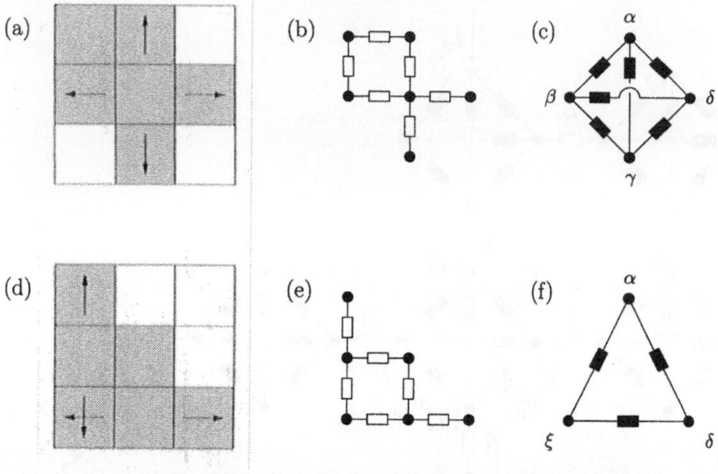

Figure 6. Generators of finitely ramified Sierpinski Carpets for four (a) and three contact points (d), their interpretation as a generator network (b+e) and the transformation into a rhomboid (c) or triangular network (f).

on large length scales the fractal becomes isotropic. This restoration of isotropy on fractals is already known [12].

After enough iterations, each effective resistance changes with a unique factor λ from one iteration step to the next. Thus the resistance scaling exponent ζ is given by

$$\zeta = \frac{\ln(\lambda)}{\ln(n)}.$$

The connecting resistances $\hat{R}$ between the rhombi (figure 7) can be neglected which leads to a different function $\mathbf{R}' = g(\mathbf{R})$. Since the effective resistances in the rhombus grow indefinitely during the iteration steps, the influence of the connecting resistances $\hat{R}$ decreases. We find that iterations of both functions converge to the same value. Thus either $f(\mathbf{R})$ or $g(\mathbf{R})$ can be used to calculate λ.

The basic idea of our algorithm is to iterate the function $f(\mathbf{R})$ (or $g(\mathbf{R})$). Unfortunately, only in a few simple cases it is possible to find explicit forms for these two functions.

To circumvent this drawback we do not deal with these functions themselves, but use the current-voltage relations between the contact points of the rhombus for the iteration. After each iteration step we calculate the effective resistances and their scaling factors. Therefore we do not have to find $f(\mathbf{R})$ (or $g(\mathbf{R})$) explicitly but iterate them in a very implicit way. A detailed description of the iteration and the corresponding MATHEMATICA program is given in [13].

The resistance scaling algorithm gives the random walk dimension with arbitrary accuracy. Hence this algorithm is a powerful tool to investigate the random

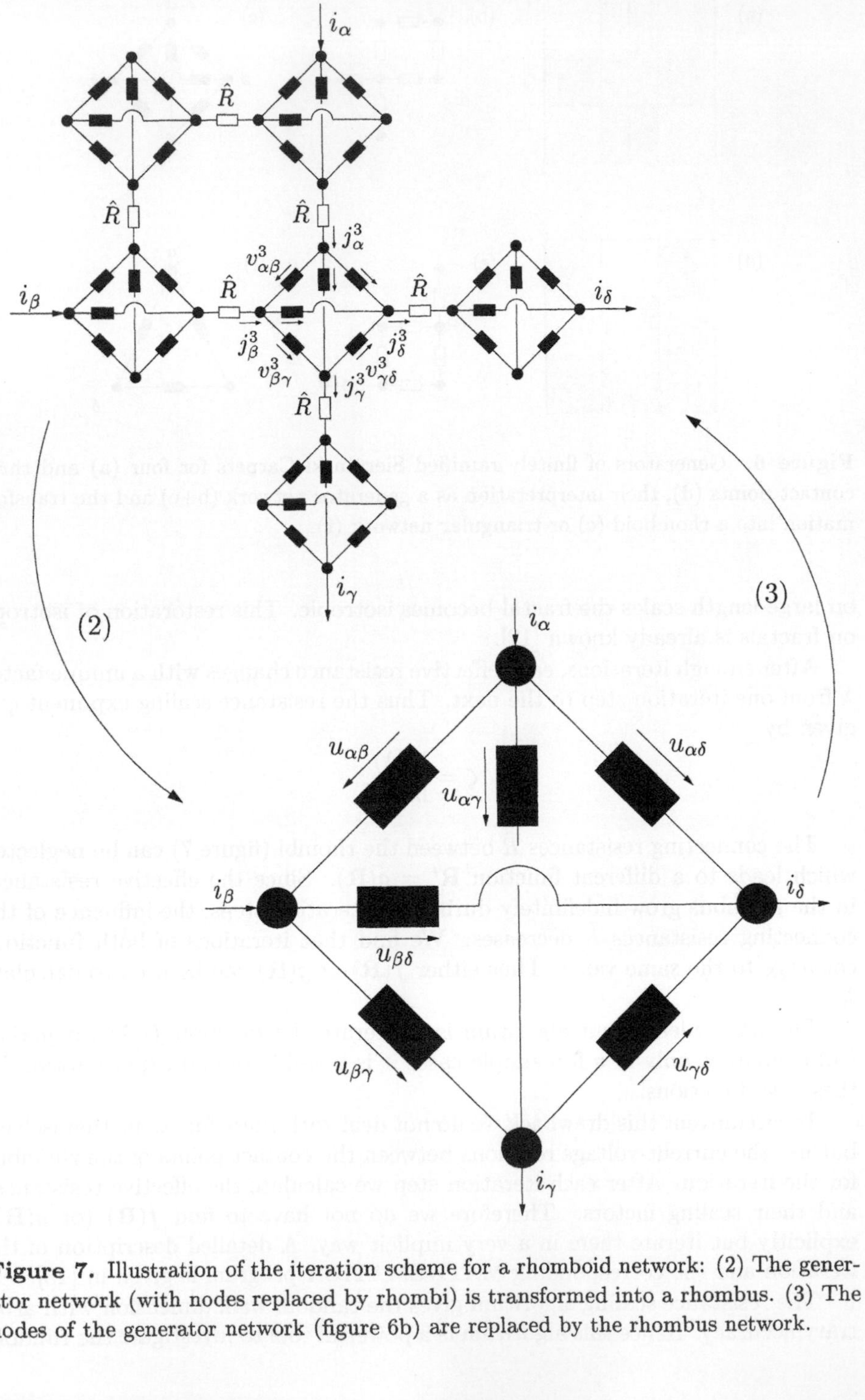

Figure 7. Illustration of the iteration scheme for a rhomboid network: (2) The generator network (with nodes replaced by rhombi) is transformed into a rhombus. (3) The nodes of the generator network (figure 6b) are replaced by the rhombus network.

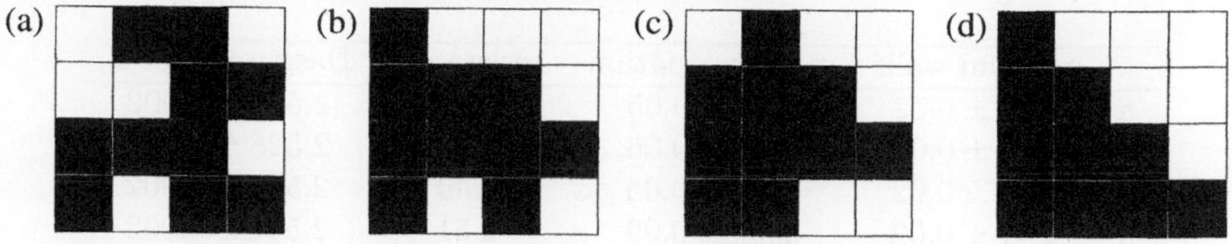

Figure 8. Four examples of Sierpinski carpet generators.

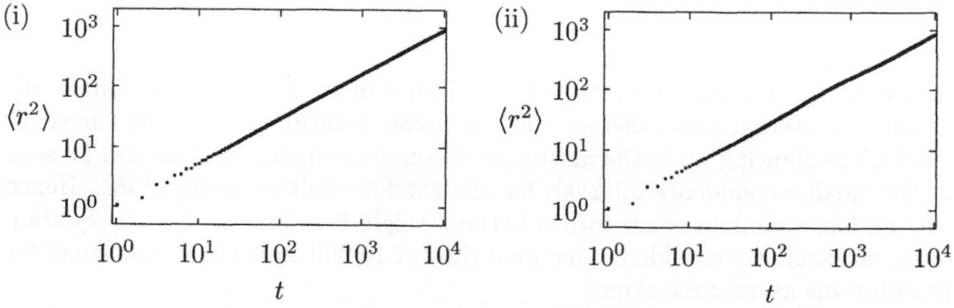

Figure 9. Log-log-plots of $\langle r^2 \rangle$ over t for the random walk algorithm (i) and the master equation iteration (ii) for carpet pattern shown in figure 8a.

walk dimension of finitely ramified fractals.

5 Test results and conclusions

We applied all three methods explained in the previous sections to get estimates for the random walk dimension of the four 4×4 carpets investigated in [4] (see figure 8). The random walks and the iteration of the master equation are done for 10,000 time steps, furthermore we averaged over 10,000,000 walkers. The resulting log-log-plots for the carpet pattern given in figure 8a are shown in figure 9.

To approximate the random walk dimension by linear regression we choose the points with times 16, 32, 64, 128, 256, 512, 1024, 2048, 4096 and 8192, since these points have constant distance on the logarithmic scale. The d_w values resulting from the regression together with confidence intervals of 95% are shown in table 5, as well as the theoretical values resulting from the resistance scaling method. For comparison in the last column of table 5 the values of [4] are cited.

Since the resistance scaling method allows to compute the random walk dimension with arbitrary accuracy, these values can be taken as reference values. As can be seen from table 5, the values for d_w can be really different although all four example carpets have the same fractal dimension. Let us remark that the random walkers choose different starting points, whereas the starting point

	random walk	master equation	resistance	Dasgupta et. al.
a	2.68 ± 0.02	2.71 ± 0.05	2.66	2.538 ± 0.002
b	2.52 ± 0.02	2.52 ± 0.06	2.58	2.528 ± 0.002
c	2.47 ± 0.02	2.49 ± 0.05	2.49	2.524 ± 0.002
d	2.47 ± 0.02	2.50 ± 0.09	2.51	2.514 ± 0.002

Table 5. Estimates for d_w for the four carpet patterns of figure 8 resulting from our three methods and from [4].

for the master equation method is fixed. Hence in the random walk data additionally an average over different starting points is included, reducing the small oscillations showing up in the master equation data (figure 9). This can be seen in the smaller confidence intervals for the random walk estimate of d_w. Hence the random walk data points form a better straight line than the master equation data, although the considered maximal time of 10,000 seems to be too small for reaching the asymptotic slope.

This can be confirmed with the resistance scaling method. In the algorithm shown in section 4 the number of iterations needed until convergence is reached gives an estimate of the size of the carpet which has to be covered by random walkers until the asymptotic behaviour can be seen. For the carpet pattern shown in figure 8a for instance 130 iterations are necessary. Assuming a random walk on a graph with edge lengths in the range of an atom radius, say $1\text{Å}=10^{-10}\text{m}$, the whole carpet should have a size of $4^{130} \cdot 10^{-10}\text{m}=1.8 \cdot 10^{68}\text{m}=2 \cdot 10^{52}\text{ly}$. These are cosmic dimensions which of course can never be reached in reasonable computing time.

Hence the resistance scaling algorithm is the most accurate one presented here to get an estimate of the random walk dimension. But the random walk algorithm and the master equation method yield additional information about the probability distribution of the walkers after a certain time. Thus these algorithms may be used to get some more insight in the scaling behaviour of the probability distribution for random walks on fractals.

Bibliography

1. Havlin, S. and Ben-Avraham, D. (1987). Diffusion in disordered media. *Advances in Physics*, *36*, 695–798.

2. Bunde, A. and Havlin, S., editors (1996). *Fractals and disordered systems.* Springer, Berlin.

3. Falconer, K. J. (1997). *Techniques in fractal geometry.* John Wiley & Sons Ltd, Chichester.

4. Dasgupta, R., Ballabh, T. K. and Tarafdar, S. (1999). Scaling exponents for random walks on Sierpinski carpets and number of distinct sites visited: A new algorithm for infinite fractal lattices. *Journal of Physics A: Mathematical and General, 32*, 6503–16.

5. Seeger, S., Franz, A., Schulzky, C. and Hoffmann, K. H. (2000). Random walks on finitely ramified Sierpinski carpets. Accepted for publication in *Computer Physics Communications.*

6. Franz, A., Schulzky, C., Seeger, S. and Hoffmann, K. H. (2000). An efficient implementation of the exact enumeration method for random walks on Sierpinski carpets. *Fractals, 8*, 155–61.

7. Tetali, P. (1991). Random walks and the effective resistance of networks. *Journal of Theoretical Probability, 4*, 101–9.

8. Palacios, J. L. and Tetali, P. (1996). A note on expected hitting times for birth and death chains. *Statistics & Probability Letters, 30*, 119-25.

9. Given, J. A. and Mandelbrot, B. B. (1983). Diffusion on fractal lattices and the fractal Einstein relation. *Journal of Physics B: Atomic, Molecular and Optical Physics, 16*, L565–9.

10. Barlow, M. T., Bass, R. F. and Sherwood, J. D. (1990). Resistance and spectral dimension of Sierpinski carpets. *Journal of Physics A: Mathematical and General, 23*, L253–8.

11. Franz, A., Schulzky, C. and Hoffmann, K. H. (2000). The Einstein relation for finitely ramified Sierpinski carpets. Submitted to *Nonlinearity.*

12. Barlow, M. T., Hattori, K., Hattori, T. and Watanabe, H. (1995). Restoration of isotropy on fractals. *Physical Review Letters, 75*, 3042–5.

13. Schulzky, C., Franz, A. and Hoffmann, K. H. (2000). Resistance scaling and random walk dimensions for finitely ramified Sierpinski carpets. Accepted for publication in *SIGSAM Bulletin.*

Why study financial time series?

M.D. London, A.K. Evans and M.J. Turner

Institute of Simulation Sciences, SERC, Hawthorn Building, De Montfort University, Leicester LE1 9BH

Abstract

This article gives an introduction to quantitative finance: why financial time series are interesting and what we stand to gain by studying them. The first part motivates and describes the role that complex systems, self-organised criticality and universality might play in unravelling the mystery surrounding the stylised and non-trivial empirical regularities observed in markets. The second part of the paper then describes each of these empirical regularities in turn with some mathematical background and popular statistical models.

1 Stock prices and stochastic time series

The erratic nature of financial processes such as stock market and exchange rate prices is an inescapable feature of modern life. Prices are now quoted on the hourly news and affect us all in some way. Figure 1 shows a graph of an American stock market index, the New York Average (NYA) with daily observations from 1980-89. One event clearly stands out from the rest: the crash of 1987 that left many economists and investors puzzled and dejected [8, 75].

A mathematician might describe the time series in figure 1 as stochastic or random; another word for erratic. In recent decades stochastic time series (or signals) and therefore stochastic modelling have become an increasingly important part of the mathematical sciences due to the number of naturally occurring systems in which stochastic signals are inherent. Examples of such systems include traffic, neural activity, heart rate variability, diffusion, earthquakes, forest fires, speciation and extinction, light emission from quasars and countless more (see for example [5, 79, 69, 12, 14]).

In economies the behaviour of stock prices, including apparent anomalies like market crashes, *emerges* as the result of a lot of complicated interactions. Interactions between companies, financial agents, governments and just chance [80]. Emergence and complexity are words, often used synonymously, that describe a new concept of systems whose behaviour cannot be understood in a reductionist manner [5]. 'Complex systems' are systems with many (often identical) interacting parts, in which simple rules (which specify how the system evolves) lead to complicated and unpredictable behaviour. Thus complexity, like chaos ([69] and

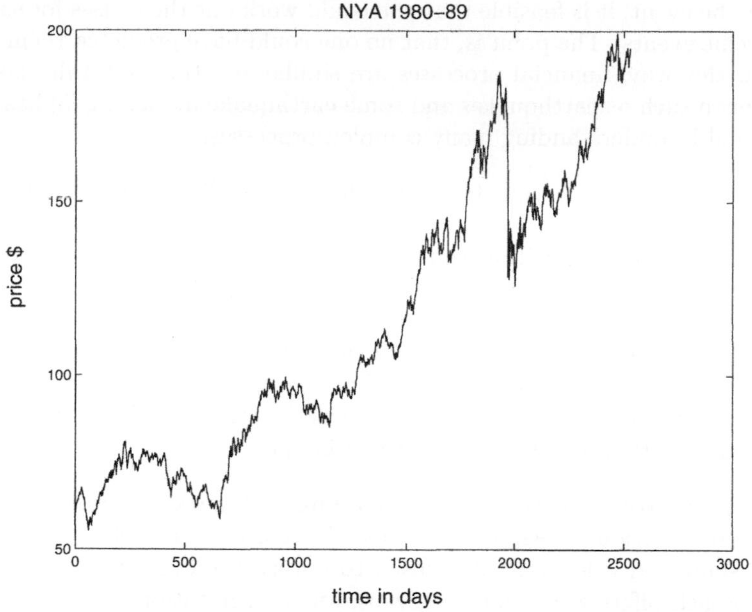

Figure 1. The New York Average from 1980-89, showing the crash of late 1987 at around day 2000.

defined formally below), has a mathematical definition that is not the same as its English language definition and which is often twofold. It essentially describes the variation within a structure formed by a complex system or the degree to which such variation is present. We might intuitively associate this with the entropy of the system but the two concepts are not the same. Entropy, which describes the amount of disorder (or randomness) in the system, is not maximised in systems with high complexity. Complexity is a more interesting kind of variation and one that is difficult to quantify. For example the landscape of the Earth has a high degree of complexity, but its entropy is far from maximum. In fact, the earth would look very boring if its entropy were maximum.

Understanding these microscopic interactions in complex systems turns out to be quite difficult. The most important reason for this is that the main way in which we test our models is to compare synthesised data with macroscopic empirical data: data in which all the important microscopic workings have already taken place and been realised in such a way that we cannot easily work backwards to see what they were. In finance we are often left guessing at what series of events could have led to that bubble or crash. Seismologists are often left wondering what series of small tectonic movements could have initiated that large earthquake, and motorists are often frequently bemused by that frustrating phenomenon of the traffic suddenly coming to a violent halt only to clear again straight away revealing a clear road.

After the event, it is feasible that one could work out the causes for such rare and extreme events. The point is, that no one could have predicted them beforehand. In this way, financial processes are similar to other naturally occurring phenomenon such as earthquakes and some earthquake modelling [5] has proven to be useful in understanding many complex processes.

There are two main ways to approach the modelling process: the first is the imitation approach. That is, to study the statistics of the real data and try to imitate the data within some mathematical framework; to understand the process by being able to re-create it statistically. Analogies may be drawn between the constituent parts of the financial process and the mathematical mechanisms that synthetically mimic the process and this may lead to some new insight. Sometimes a model may be justified by means of some other argument based on assumptions about real markets as is the case with the Efficient Market Hypothesis (EMH, which will be described later).

The alternative is to try to formulate a financial market microcosm in which the desired behaviour 'emerges' as it does in real life. To achieve this, appropriately simplified rules must be chosen to govern the evolution of the system. This approach offers a natural framework for interpretation, but there are two obstacles to overcome: firstly, rules must be selected that give you all of the right behaviour (reproduce the statistics) and secondly, the rules must be simple enough to make mathematical analysis of the model possible. It is often surprisingly difficult to analytically show why the simplest of models behave the way they do.

Some might argue that a further difficulty in finance is that agents of the economies differ greatly in ability, importance and influence. Or that psychology adds a layer of complexity to modelling financial processes that is impossible to overcome because people in general do not behave in a consistent manner. Or perhaps that rare events like the death of Princess Diana or the recent terrorism in America cannot be foreseen and yet have an important impact on many financial processes. We argue here that the magnitudes of such events are part of a continuum, ranging from the everyday least significant events to the rarest and most devastating. With their great variety and number, these events may be governed by some universal laws. We argue that if statistical regularity exists, it may be so because the large events are related to the small events by some natural mechanism. This is a big claim and one that will need to be defended carefully. We begin by building a diverse coalition of examples of systems in which similar claims have been justified.

Modern financial research is linked to work in a diverse range of disciplines including seismology, psychology, biological physics and electronics to list but a few. Many useful metaphors are becoming available and hence a better understanding of the general principles that underly such processes; for example chaos. This brings us to the concept of universality.

2 Universality

Separating the general from the particular is a big part of what science is about [29]. A universality describes a situation in which surprisingly many particulars can be ignored and a general result still prevails. More precisely, a universality class is a class of systems whose macroscopic behaviours are qualitatively the same although the details of their construction may differ considerably. This has applications in chaos theory [69], statistical physics [12] and other fields.

For example, a system in which large samples of independent random variates are taken from any distribution and added together is governed by the central limit theorem (CLT) (discussed again later) and the resulting behaviour of the system can be described by the class of L-stable or Lévy distributions of which the Gaussian distribution is a particular case. This is one of the most celebrated universality classes. Many natural phenomenon follow this distribution: the heights of humans, the error after the best model has been fit to a data set, diffusion in disordered media [12]: any system in which there is an element of chance potentially obeys this law.

Another example is the Fibonacci sequence and the golden ratio [43]. The sequence begins by taking two numbers a_1 and a_2 and creating the rest of the sequence using the rule, $a_{n+1} = a_n + a_{n-1}$. For example: a={1,1,2,3,5,8,13,21...}. It can easily be shown that the ratio, a_{n+1}/a_n, tends to, $(1 + \sqrt{5})/2 \approx 1.618$ regardless of the choice of a_1 and a_2. Much of nature's geometry is governed by these so called golden ratios [20]. For example, shell spirals, the branching of plants, the number of petals on different types of flowers, the patterns of seed heads and pine cones, the arrangement of leaves and many ratios of measurements of the human body [43]. There is evidence to suggest that building things with these ratios in mind is a very efficient way of doing things [23]. Furthermore, this fact is now being reflected by human taste, from art to music to the human face, it appears that we find geometries that obey this ratio pleasing to the eye. Some work (including [71] and an 'art-house' film called 'π') has argued that the golden ratio appears in financial time series though this is very difficult to verify and not generally accepted.

Chaos theory, another example, is a great achievement in the understanding of the mathematical principles at work in our universe. It shows us how under certain conditions, simple systems can behave unpredictably. A system, f, defined by, $x_{t+1} = f(x_t))$, for some nonlinear function f, is chaotic when initially close orbits diverge exponentially, $|f_t(x_0) - f_t(x_0 + \epsilon)| \sim \exp(\lambda t)$, or more precisely that the average value of this measure λ is positive [7, 69],

$$\lambda(x_0) = \lim_{n \to \infty} \frac{1}{n} \sum_{k=1}^{n} \ln f'(x_{k-1}) \tag{2.1}$$

assuming that f' exists. Although there are many types of system that exhibit chaos, if we know that they are chaotic then we can expect certain regularities, particularly concerning the route to chaos for example, period doubling and windows of order governed by universal ratios [69]. We have a good understanding of how chaos works and can thus now explain why it is difficult to predict so many natural systems like weather [55, 76], population dynamics [63], fluid mechanics [49] and healthy heart rate variability [47].

Fractal geometry [58] is another example of a ubiquitous phenomenon which can be seen and embraced given a change of thinking. Often the origin of fractals is linked with chaos, since chaotic orbits may create fractal objects in some phase space [7]. However, many systems and mechanisms that produce fractals, such as Diffusion Limited Aggregation (DLA) [21] and non-chaotic complex systems [5, 14], are known not to be chaotic indicating that we do not have the complete picture as yet. One promising new mechanism known as criticality [5] might be one of the missing pieces of the puzzle.

Criticality is currently less well understood. It reveals itself through power law probability distributions with infinite second moments, fractal noise and Zipf's law [5]. A simple example recently quoted is known as 'potato logic' [5, 14]. If one were to throw frozen potatoes at a wall repeatedly and then carefully weigh all of the fragments on the surrounding floor, then it would soon become clear that the size and frequency of the fragments are related by the simple power law, $f \propto 1/\text{size}^{\alpha}$, $\alpha = 2$ (actual experiments carried out by [67]). There is no obvious or intuitive reason that this should be so. What's more, in [67] the experiment was repeated with chunks of gypsum, soap and other substances, the results were also power laws. Similarly earthquakes, traffic jams, forest fires and many more natural systems obey this same power law with exponents $1 < \alpha < 3$ [5, 14]. This is one of the new hot topics in mathematics.

The above examples illustrate that understanding universality is about understanding certain mathematical mechanisms. For instance stability under some mathematical operation (like addition, or convolution), or the nature of some condition like equation (2.1). Understanding such mechanisms will explain why certain details that don't affect the universality class that the system falls into. For example, in the case of the (Gaussian) CLT, one can see the unimportance of the moments of order greater than two by looking at the limiting form of the characteristic function (Fourier transform of the PDF). (see appendix 13.)

The important thing about universality is that all we have to do to understand a whole (universality) class of processes is to invent the simplest 'toy' model which contains the relevant basic features necessary to belong to the class. We can then use this simplest model to understand and make inference about the whole class of processes.

Universality is relevant in financial modelling for two reasons. Firstly different markets trading different financial instruments in different countries have certain dynamical features in common, and they are not trivial (examples of universality classes already understood). These features will be discussed in the latter half of this paper. Secondly, other natural and social systems exhibit similar dynamical behaviour as well. This tempts us to consider the possibility that some very general principles are at work. The exposition of such a universality class to describe financial processes is the holy grail of financial research. Before we do so, we go back one hundred years to the first attempts to justify models of price processes.

3 A starting point: the random walk

In 1900 a French mathematician by the name of Louis Bachelier submitted a PhD thesis titled, 'The Theory of Speculation' in which he proposed a random walk model for asset prices [2]. He concluded that the price of a commodity today is the best estimate of its price in the future. The random behaviour of commodity prices was again noted by Working [81] in an analysis of time series data. Kendall [41] attempted to find periodic cycles in indices of security and commodity prices. He did not find any. Prices appeared to be yesterday's price plus some random change. He suggested that price changes were independent and that prices followed random walks. The majority of early financial research is reported [80] to have been in agreement: asset price changes are random and independent, so prices follow random walks. The random walk model is often referred to as Wiener Brownian motion (WBM) named after the Mathematician Norbet Wiener and the Scottish botanist Robert Brown. Brown [13] noticed the erratic motion of a small particle suspended in a fluid and Wiener [62] later provided a rigorous mathematical framework with which to work with what we now know as stochastic processes.

Bachelier also noticed that the size of price movements was proportional to the price, thus his model proposed that the log of price changes should be Gaussian distributed. This behaviour is often described mathematically by a model of the form [80],

$$\frac{dS}{S} = \sigma dX + \mu dt \tag{3.1}$$

where S is the price at time t, μ is a drift term which reflects the average rate of growth of the asset, σ is the volatility and dX is a sample from a normal distribution. There are problems with this notation and a better notation [33] is,

$$\frac{dS}{S} = \sigma N(t)(dt)^{1/2} + \mu dt \tag{3.2}$$

In other words the relative price change of an asset is equal to some random element plus some underlying trend component. More precisely this model is a log-normal random walk. The Brownian motion model has the following important properties:

1. Statistical stationarity of price increments. Samples of Brownian motion taken over equal time increments have identical moments.

2. Scaling of price. Samples of Brownian motion corresponding to different time increments can be re-scaled using a simple power laws such that they too have the same descriptive statistics (moments). For example, denoting

$B(t)$ to be a Brownian process,

$$\langle |B(t+N\tau) - B(t)|^q \rangle \sim \frac{A_q}{N^\alpha} \langle |B(t+\tau) - B(t)|^q \rangle \tag{3.3}$$

3. Independence of price changes taken between non-over lapping intervals.

3.1 Efficient markets?

It is common knowledge amongst traders and anyone with an interest in the market that the distribution of price changes has fatter tails than the Gaussian and that volatility exhibits clustering (see [26] for a good review). One only has to look at the raw data to see this (Figure 1). So why did people think that prices should follow random walks?

It is often stated that asset prices should follow random walks because of the Efficient Market Hypothesis (EMH). The EMH states that the current price of an asset fully reflects all available information relevant to it and that new information is immediately incorporated into the price. Thus in an efficient market, the modelling of asset prices is really about modelling the arrival of new information. New information must be independent and random, otherwise it would have been anticipated and would not be new. In this way, asset prices are a Markov process [33].

The EMH implies independent price increments but why should they be Gaussian distributed? Perhaps the Gaussian PDF is chosen because macroscopic price movements are presumed to be an aggregation of smaller (microscopic) ones, and this aggregation is governed by the Central Limit Theorem [12].

The EMH assumes that there is a rational and unique way to use the available information and that all agents possess this knowledge. Moreover, the EMH (very ambitiously) assumes that this chain reaction happens instantaneously. In an efficient market, only the revelation of some dramatic information can cause an extreme event, yet post-mortem analyses of crashes typically fail to (convincingly) tell us what this information must have been [48].

3.2 Price response to information.

The EMH assumes a simple relationship between the reaction of the market to information and the importance of that information. Trivial information should cause trivial price adjustments and cannot be responsible for crashes, which can only be instigated by very important information. In reality however, avalanches (bubbles and crashes) are known to occur for no good reason. Panic selling (or buying) motivated by human fear, greed and imitation are also familiar. There is an obvious feedback mechanism at work in financial systems: agents' trading decisions regarding a given financial instrument are based on (amongst other things) the price of that financial instrument. And it is the agents' collective actions that form its new price. This means that the response of the price to information is more complicated: agents respond to information and then to each others responses and so on.

To gain insight into the way information is used in markets, it is useful to know something about the trading mechanisms and tools available to agents of the economy and how they are used to manipulate risk. The following is a brief description of how risk, information, agents and trading contracts form economies.

4 Strategy: microscopic market activity

Years ago Portfolio theory focused on rates of return with the occasional caveat, 'subject to risk'. Modern Portfolio theory assumes that there must be a trade off between risk and return. A simple argument outline in [80] illustrates why this should be so.

We must first assume the existence of a (virtually) risk free investment, for example, depositing money in a reputable bank. Suppose now that we could beat the bank by investing in equities without taking any extra risk. Clearly no one would invest any money in the bank. What would the bank do about that? It would have to raise its interest rates, to attract more investment in the bank, to make more money from its own lending to pay the interest on the accounts in credit and to make it more difficult for a person to make money by borrowing from the bank and investing it in equities. (There are transaction costs and differences between buying and selling prices but these do not really affect this argument.) The market is full of arbitrageurs whose job it is to seek out irregularities and mispricings and profit from them. The concept of arbitrage is important in finance [80]. Often stated as, 'there's no such thing as a free lunch', it means that opportunities to make a risk free profit cannot exist for very long before prices move to eliminate them. Thus investing in non-safe investments is about speculation.

As markets have grown and evolved, new trading contracts have emerged called derivatives [80], which use various clever tricks to manipulate risk. The value of a derivative is derived from, but not the same as, some underlying asset or price index. These special deals really just increase the number of moves financial agents have at their disposal to ensure that the better players win. Anyone that has played a game like noughts and crosses or draughts a large number of times will soon reach the stage where each game is the same and a stalemate. If you think you are intellectually superior and you want to take advantage of it then you should suggest a game of chess.

There are many different kinds of derivative traded on the world's markets today [80]. To illustrate the basic principles we introduce the simplest and most common one: the option. An option is the right (but not the obligation) to buy (a **call**) or sell (a **put**) a financial instrument at a prescribed time in the future (the expiry date) at a given price, known as the strike price, or exercise price.

Options may be used as a more pure means of speculating. If the expiry date arrives and the price has not gone the way you speculated then you simply do not exercise the option. All you have lost is the initial premium that you paid for the option; you are only paying for the right to speculate. If volatility is high then the potential gains are unbounded but the loss is always limited to the initial premium paid. Thus, the value of an option is a function of volatility, the strike price, and the time left till expiry.

Options may also be used as a means of hedging, that is as an insurance policy. Suppose an investor owns shares in a particular company. If he thinks there is a chance that the share price may take a dive, but he doesn't want to risk liquidating his stock, then he may wish to buy some *put* options in the same company. If the price goes down, he has the right to sell his stock at a price (hopefully) higher than the current price. He can then buy it straight back generating a profit and his long term investment resumes. If the price does not take a dive then the investor simply carries on with his long term investment losing the premium paid on the options but profiting from having left his long term investment going.

A sensible question to ask at this point is: how much should one pay for an option? Before the introduction of the Black-Scholes (BS) formula [9] in 1973, options were valued quite subjectively. The Black-Scholes formula, which is an algebraic equation (derived from a partial differential equation), assumes that the price of the share follows a log-normal random walk and works by taking advantage of the fact that the value of the option and the value of the underlying asset are correlated. For further reading on the mathematics of financial derivatives and a derivation of the Black-Scholes equation see [53, 80].

It would seem that the way to use the BS formula is to estimate the parameters, the interest rate, the exercise price of option, time till expiry, the price of the underlying and the volatility, substitute them into the formula and then estimate the value of the derivative product. It turns out that this is no longer the most common use of option models. This is partly due to the fact that, in practice, it is difficult to measure the volatility of the underlying asset. However, despite these difficulties, option prices are still quoted in the market. This suggests that, even if we do not know the volatility, the market does. So, having substituted into the formula, the interest rate, the exercise price, the time till expiry and the price of the underlying (all of which are easy to measure), all that remains is the derivative price and the volatility. Since (according to the model) there is a one-to-one correspondence between the volatility and the option price, we could equally well take the option price quoted on the market and substitute that in instead to estimate the market's opinion of the volatility over the remaining life of the option. This estimation of the volatility from the BS formula and the price of the option is called the *implied volatility* [80, 18].

A common empirical feature of the implied volatility is that it isn't constant across the exercise prices [80, 18]. That is, if the other parameters of the model are fixed, then the prices of options across the exercise prices should reflect a uniform value for the volatility. The volatility of options with exercise prices far from the current underlying price (so called in-the-money options) is often greater than those with exercise price close to the current price (at-the-money). This is known as the volatility 'smile', there are other less common possibilities as well such as the volatility 'frown' [80, 18].

5 Adaptive markets

The BS model has many such shortcomings, including the major failure of assumption involving the distribution for the price increments of the asset. These shortcomings will be discussed in detail in the second half of this paper. Jessica James [37] (head of research in the strategic risk management advisory group at the first bank of Chicago, London) explains that the Black-Scholes formula has not lost popularity since its shortcomings (assumption of log-normal distribution) became apparent, because of the way in which it is used. The Black-Scholes volatility and the price of an option are now so closely linked in the market that an option is usually quoted in option volatilities or 'vols' (which are displayed on traders screens all across the world). Traders know that market distributions are not log-normal, so they adjust the option price to take account of this. Traders make this adjustment by allowing the estimated volatility, which determines the price of the option, to vary with the options strike price as described above. This variation can have a number of characteristic shapes. This crucial piece of flexibility removes at a stroke, most of the limitations of the model, and allows options to be valued based on a variety of estimated future price distributions. Common variations are the 'vol smile', 'frown' and 'smirk'.

A paper by Potters, Cont and Bouchaud [72] verifies, by studying the prices of options on liquid markets that the market has empirically corrected the Black-Scholes formula to account for the 'fat tails' and correlations in the scale of fluctuations. Potters *et al* suggest a replacement of Gaussian price increments with increments that are the product of two random variables, one of which contains time correlations.

This highlights one reason why the economy is so fascinating. It is greater than the sum of its parts. It is of course driven by human beings buying and selling assets and other financial instruments, but these people do not decide and cannot comprehend the affect they are collectively going to have on the economy (they wish they could). The observed universal characteristics of markets that emerge through human activity do not seem understandable through the reductionist approach of observing the actions of individuals.

6 Empirical regularities

The remainder of this paper is devoted purely to exposing facts and does not focus on any theory of how those facts might have come about. Its purpose is to establish empirical regularities, and suggest known mathematical distributions and mechanisms that are consistent with observation. We will use several real life data sets [1] to make inferences about financial time series in general. These

Price Series	Acronym	Period	Size	Frequency
Dow Jones Industrial Average	DJIA14	1987-00	50000	14 obs/day
Financial Times Industrial Index	FTIID	1933-00	17936	Daily
New York Average	NYA	1960-89	8339	Daily
New Zealand forty share	NZ40D	1970-00	7895	Daily
Japanese topix	TOPXD	1950-00	13948	Daily
French Franc/US$	FRF	1971-00	7408	Daily
Belgian Franc/US$	BEG	"	"	"
Japanese Yen/US$	JPY	"	"	"
Netherlands Guilder/US$	NLG	"	"	"

Table 1. Details of the data sets used throughout this thesis.

data sets are described in Table 1 and will often be referred to thereafter by their acronyms.

7 De-trending the data

The raw price data that we analyse contains a trend component. A graph of any asset price time series over a long time looks a bit like $y = e^x$. It always starts in the bottom left corner of the graph and ends up in the top right and grows exponentially, like the top left graph in figure 2. We are not interested in studying this feature of the data. Exchange rate time series do not display such underlying trends but the following de-trending process proposed for the asset price series can safely be applied to exchange rate data sets without undesirable side affects. The way in which the trend is removed is as follows.

Let $p(t)$ denote the price at time t and let $x_T(t)$ denote the de-trended log price increment over a sampling interval T measured in days,

$$x_T(t) = \log p(t) - (a + b \log p(t-T)) \approx \log p(t) - \log(t-T), \qquad (7.1)$$

where a and b are constants that describe the line of auto-regression. Formally, $x_T(t)$ are the Auto-Regressive 1 (AR1) residuals [70] of the log of the price, though in practice for daily data there is not much difference between this and the log increment of the price. Taking the AR1 residuals is more important when analysing higher frequency data in which there is more serial correlation.

Figure 3 shows the de-trended log price increments of the NYA from 1960-89 and Gaussian random numbers. Both sets have been normalised for comparison (and shifted vertically so we can see them separately). Before any analysis is carried out, it is clear that rare events happen more often than they would if the data were Gaussian and that the large and small events seem to cluster together. This is the first clue that a simple random walk will not be sufficient to explain

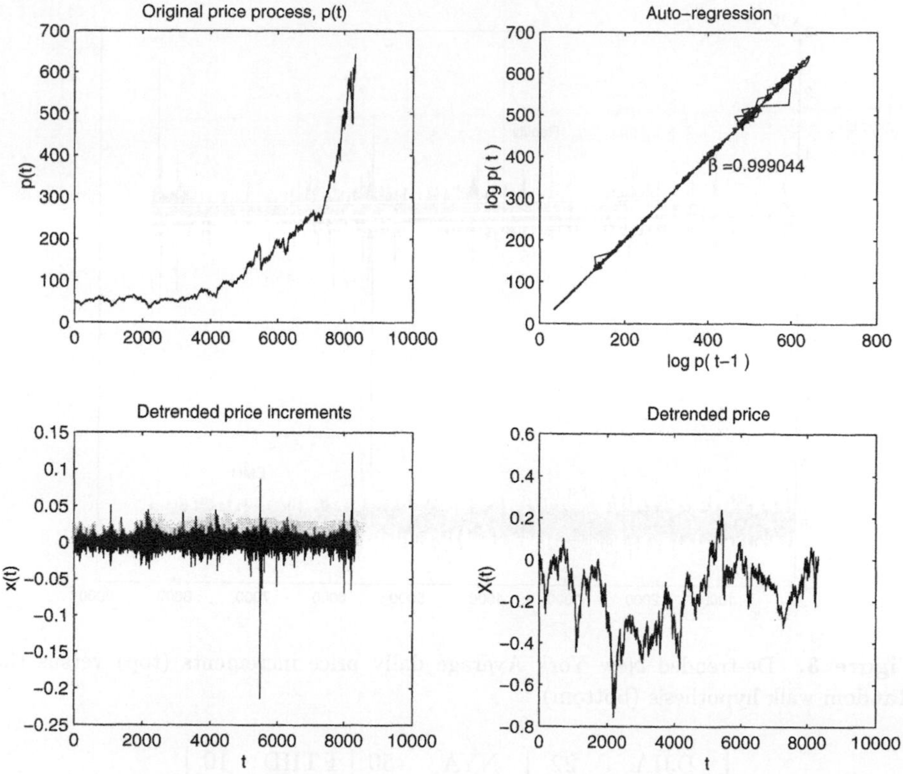

Figure 2. Illustration of the detrending process. Data used in this example is the NYA. $p(t)$ is the price at time t time steps in days, $x(t)$ is the de-trended increments and $X(t) = \sum^{t} x_{t'}$.

all features of financial time series. In the following sections we scrutinize the data and catalogue the empirical facts.

8 Probability Density Function (PDF)

For processes that follow a log-normal random walk, increments $x_T(t)$ will be Gaussian distributed on all scales. Figure 4 shows the probability density function (PDF) on three time scales: daily, weekly and monthly, for the Dow Jones Industrial Average (DJIA) index and the Financial Times Industrial Index (FTIID). The figures also show the Gaussian best fit as the dashed curve. The distributions have a sharply peaked mean and 'fat tails' as compared to the Gaussian. A distribution that has these characteristics is mathematically referred to as being leptokurtic and the extent to which this condition occurs is quantified by a measure called the kurtosis defined as,

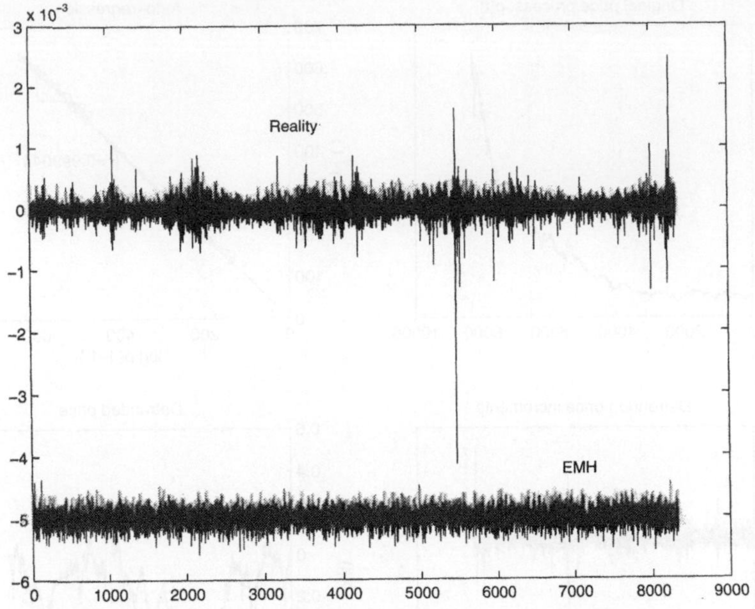

Figure 3. De-trended New York Average daily price increments (top) versus the Random walk hypothesis (bottom).

DJIA	22	NYA	50	FTIID	10
TOPXD	182	NZ40D	47	FRF	7
BEF	5	NLG	5	JPY	13

Table 2. Kurtosis for the daily increments of the datasets.

$$\kappa = \frac{\langle (x-\mu)^4 \rangle}{{\mu_2'}^2} - 3, \tag{8.1}$$

where μ and μ_2' are the expected value of x and x^2 respectively and $\langle . \rangle$ denotes the averaging operator. This measure κ is defined such that it is zero for the Gaussian.

For an independent and identically distributed (IID) process the kurtosis should scale according to, $\kappa(T) \sim 1/T$ [77], where T again, is the time scale measured in days. A study [17] has suggested that the kurtosis scales much more slowly and according to a power law, $\kappa \sim T^{-\beta}$ with $\beta \approx 0.5$. Figure 5 shows the kurtosis of $x_T(t)$ against the sampling increment size, T. The dashed curves represent the power law best fits and β (inset) is the scaling exponent for the power law. The curves fit the data but not always incredibly well. The

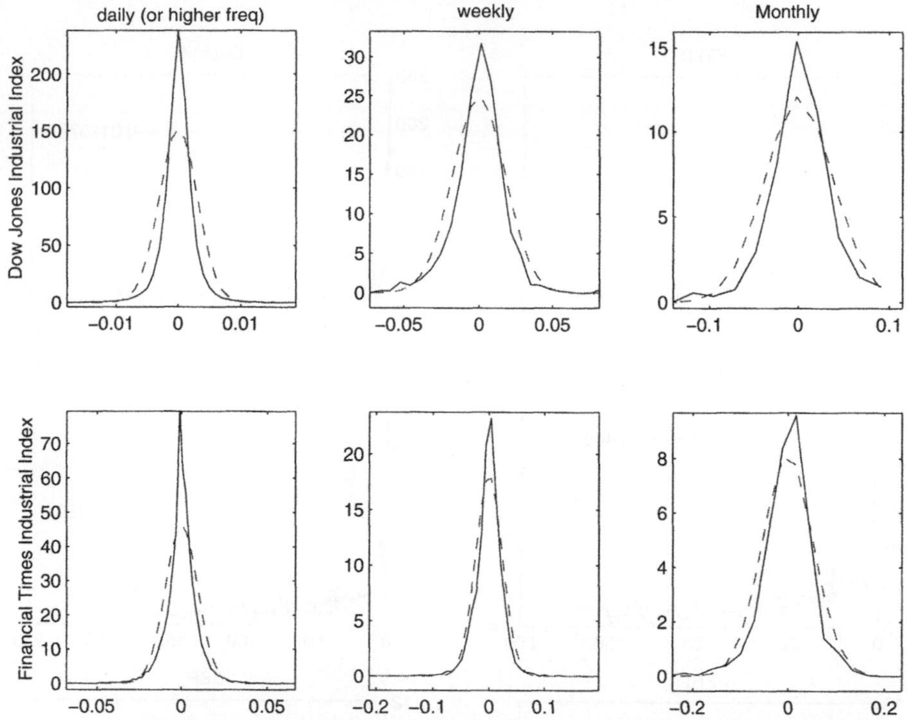

Figure 4. PDF for the de-trended Dow Jones Industrial Average (top) and the Financial Times Industrial Index (bottom) with Gaussian best fit for comparison (dashed). The horizontal axis measures the log price increments and the vertical axis is the estimated probability.

exponents are $\beta \approx 0.3$ for the intra-daily DJIA14, in the range $0.5 < \beta < 0.6$ for the daily stock market indices and in the range $0.9 < \beta < 1$ for the exchange rates. The kurtosis of the financial data does decay more slowly than an IID process, and the extent to which this happens is much greater for the stock markets than the exchange rates, particularly for the higher frequency DJIA14.

The significance of a positive kurtosis is that the likelihood of rare events is much greater than the log-normal model predicts. This observation appears to be at odds with the Efficient Market Hypothesis (EMH) because it implies that the individual contributions to a price increment (the affect of the actions instigated by the arrival of new information mentioned above in the EMH definition) are not random. There is however another possibility, that they *are* random but do not have finite variance.

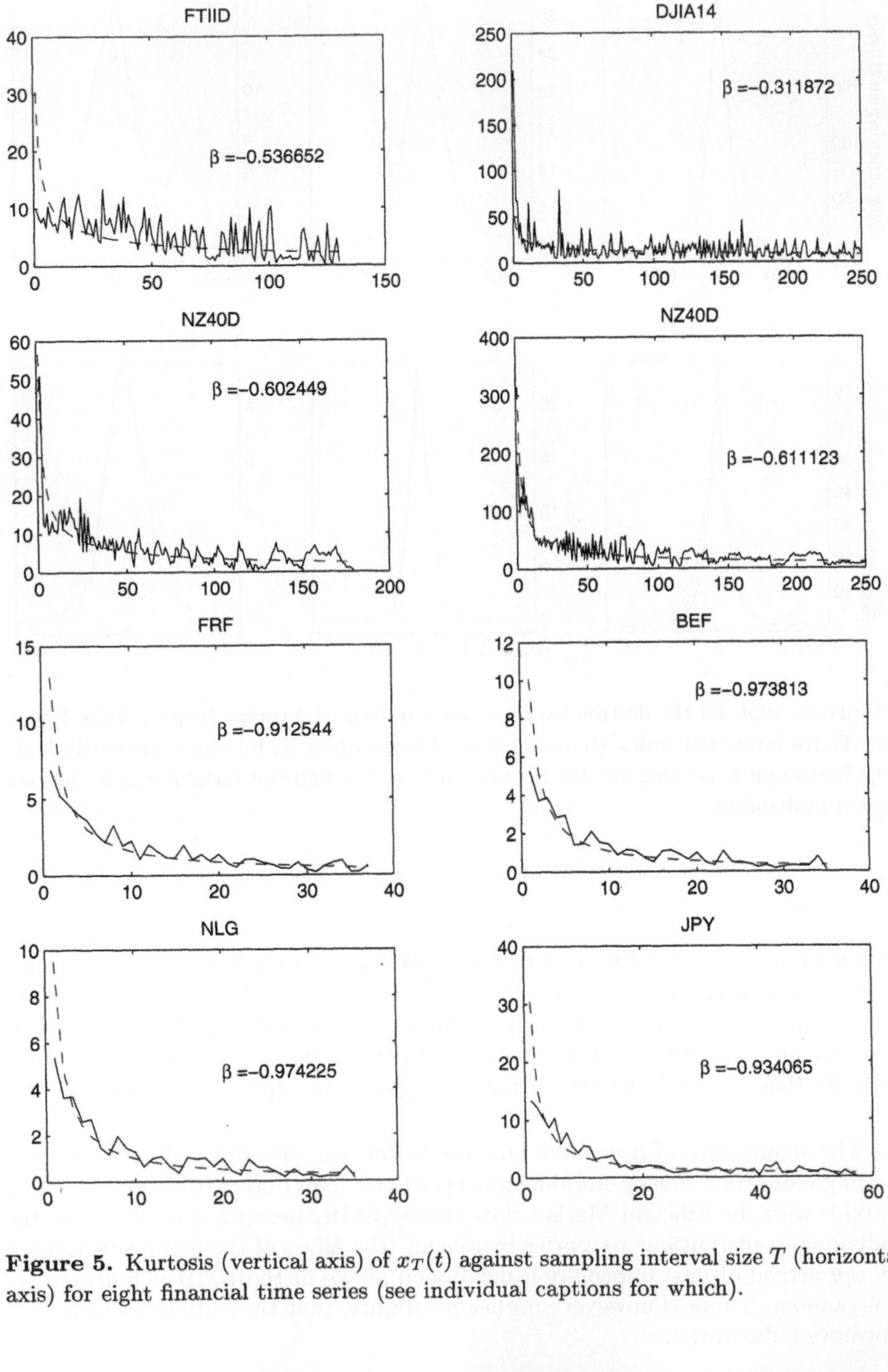

Figure 5. Kurtosis (vertical axis) of $x_T(t)$ against sampling interval size T (horizontal axis) for eight financial time series (see individual captions for which).

8.1 Stable (fractal or Lévy) distributions

There is another class of stable distributions (distributions for which the sum of two Independently and Identically Distributed (IID) random variables is identically distributed) which do not necessarily have finite variances and or means. These are commonly known as Lévy distributions [51, 78], named after Paul Lévy. It turns out that these distributions do have fatter tails and sharper peaks at the mean than the Gaussian. The distribution is normally specified through its characteristic function [77] (Fourier transform of the PDF),

$$\phi(k) = \exp(-a|k|^{\mu}), \quad 0 \leq \mu \leq 2. \tag{8.2}$$

This is because except for a few values of μ there is no closed form for the PDF (the inverse Fourier transform of $\phi(k)$). When $\mu = 0$ the PDF is a delta function, $\mu = 1$ gives the Cauchy distribution and $\mu = 2$ is the Gaussian [77]. There are also closed form expressions for $\mu = 1/2$ and $\mu = 1/3$ [12]. It can be shown [52] that this distribution has the property that its tails follow the asymptotic power law,

$$\frac{1}{|x|^{1+\mu}},$$

which is relevant from the point of view of complexity and criticality theory.

Stable distributions enjoy the self-affinity property,

$$P(x_{NT}) = \frac{1}{N^{1/\mu}} P\left(\frac{x_T}{N^{1/\mu}}\right), \tag{8.3}$$

verifying that such processes are fractal, a property which will be discussed again and again on progressively more detail throughout this chapter.

Are the increments, $x(t)$ Lévy distributed random variables? Figure 6 shows a Lévy distribution fit to the same PDF shown in Figure 4, of the DJIA14 data set. The parameter μ was 'hand picked' to make the fit near the mean good. And from the graph, the shape of the Lévy distribution is similar to the empirical distribution around the mean. On closer inspection of the tails however a discrepancy reveals itself. They are too fat: they decay more slowly than those of the empirical distribution.

Regression and iterative methods normally used to estimate the parameters of the stable distribution such as the Mc-Culloch [64] method and the Koutrouvelis [45, 46] are numerically unstable giving non-sensible answers (like infinity or not-a-number) which confirms that the overall shape of the distribution is not consistent with a stable distribution even though the centre and the tails may (separately and inconsistently) display stable distribution attributes.

Some studies [44, 34, 61] have proposed the use of a truncated Lévy flight as a model for the PDF of market returns. The parameter μ can then be chosen to fit the centre of the distribution and the multiplicative truncation factor means that they decay more like the empirical PDF, according to an equation of the form (8.4).

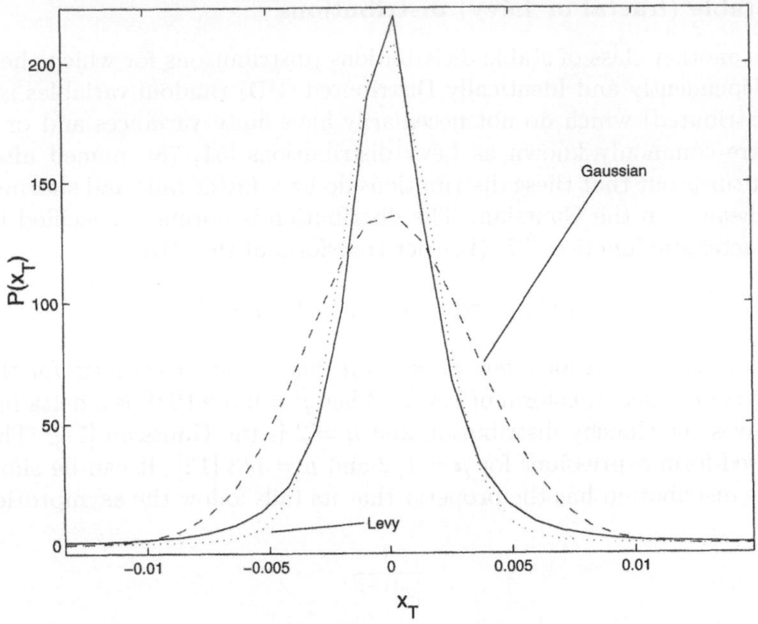

Figure 6. PDF: Lévy distributions and FTIID stock market index.

8.2 Leading order behaviour of tails

This leads us to the issue of the scaling (decaying) behaviour of the tails of the empirical PDF. Some research [11] report a power law, $p(x) \sim x^{-\beta}$, and others that the tails decay like those of the log-normal distribution ([2] and references in [19]). This issue is linked to the matter of whether the second moment of the empirical distribution is finite. In general, if the distribution is continuous and its tails decay faster than $1/x^3$ then the process's second moment exists [12, 77].

Figure 7 shows the log magnitude (of price increment) against log frequency. The curves are *nearly* straight and therefore the tails are close to following a power law though the true rate of decay is very slightly faster. The best fit lines have slopes, β, ranging from just under three to just over four and there is no noticeable difference between the stock markets and the exchange rates. The rate at which the tails of the PDF decay determines whether or not the moments of the distribution exist.

A study [17] and others have proposed that the empirical distribution is characterised by a truncated Lévy distribution whose tails would decay as,

$$p(x) \sim x^{-\alpha} \exp\left(\frac{x}{x_0}\right) \tag{8.4}$$

Another way of illustrating the likelihood of the Nth moment converging is to plot this moment for progressively larger sized samples of daily increments and

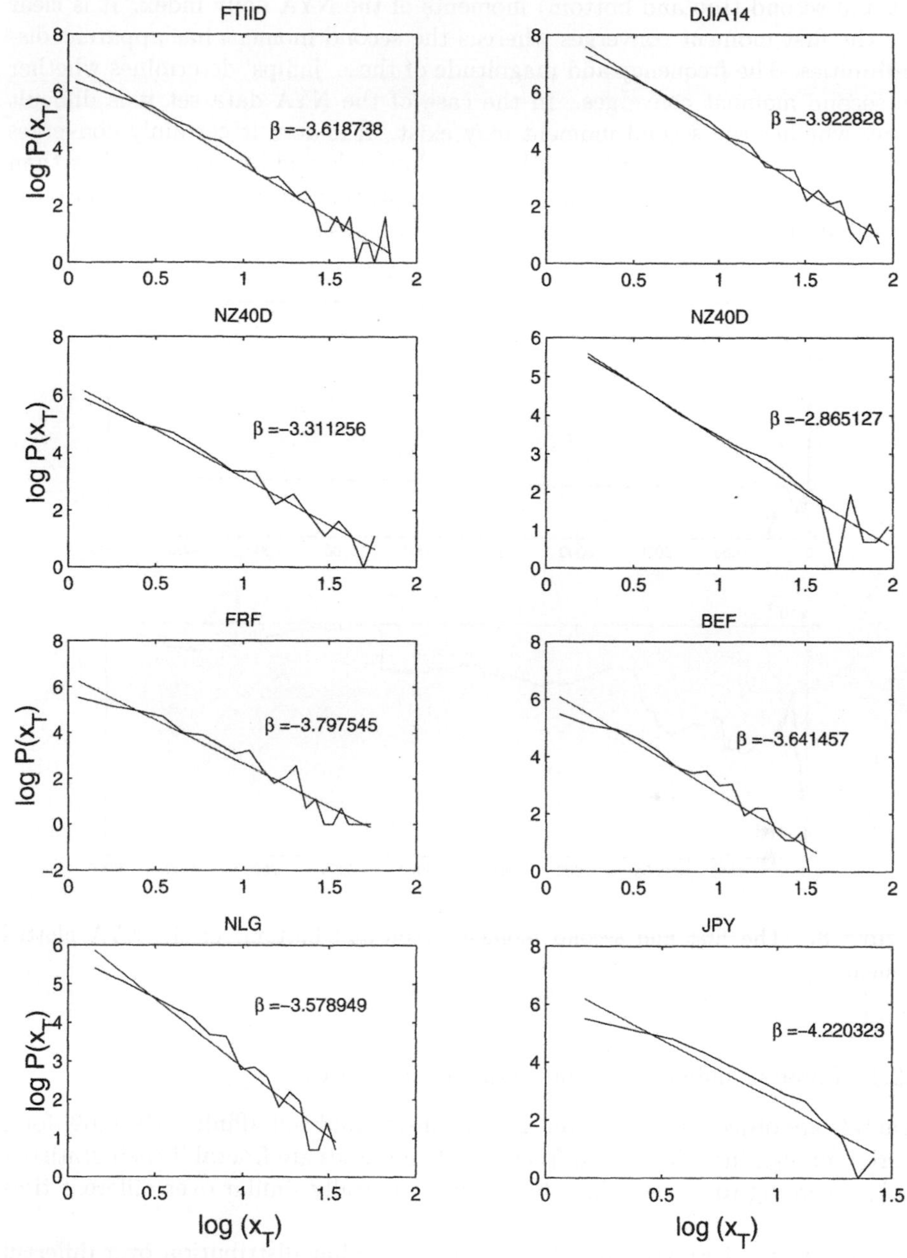

Figure 7. The tails of the PDF: log(size) against log(freq) for eight financial time series. (see individual titles.)

track its route to convergence graphically. Figure 8 shows this performed for the first and second (top and bottom) moments of the NYA daily index. It is clear that the first moment converges whereas the second moment has apparent discontinuities. The frequency and magnitude of these 'jumps' determines whether the second moment converges. In the case of the NYA data set it is difficult to say whether the second moment may exist. If it does it certainly converges very slowly. For a distribution with power law tails and an exponent less than three, these discontinuities would carry on appearing for ever and dominate the behaviour of the index.

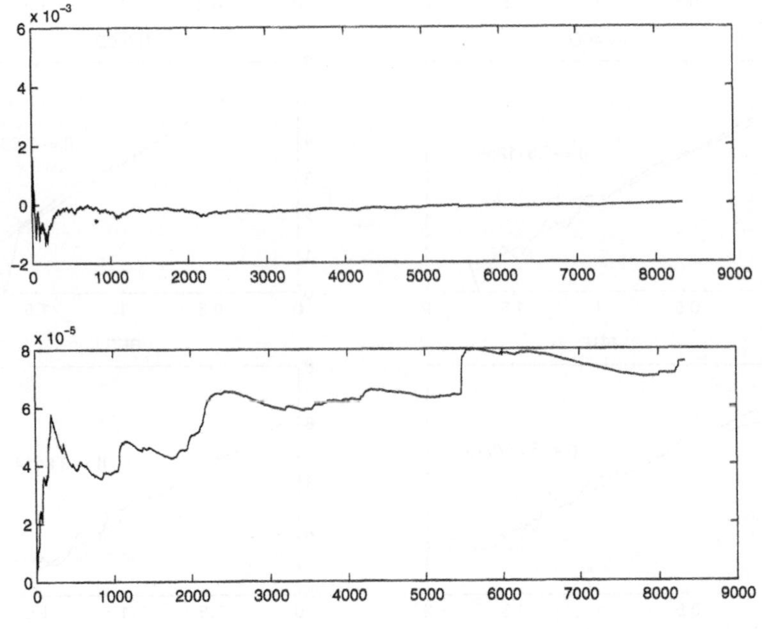

Figure 8. The first and second moments (top and bottom) of the NYA plotted sequentially.

8.2.1 Finite or infinite moments: Fractal time series

Fractals are objects with non-integer dimension and self-affinity. (See [69] for a more thorough introduction to fractals.) Time series are fractal if their statistics scale according to a power law; they are statistically similar over different time scales.

From figure 4 we see that the data has a similar distribution over different time scales. And from figures 7 and 8 that it is difficult to determine whether the second moment of the empirical distribution converges. It is generally agreed in the quantitative finance community that the second moment does converge albeit very slowly [17, 11]. A time series characterised by a PDF with an infinite (divergent) second moment has no characteristic scale and this is how it (perhaps trivially) inherits its fractal status. [70, 15, 16]. It's statistical regularity can only be expressed by means of a fractal dimension.

The fractal dimension describes two things,

1. How densely the object fills its space,
2. The structure of the object as the scale changes. For physical (or geometric) fractals, this scaling law takes place in space. A fractal time series scales statistically, in time.

The statistic usually used to describe fractal time series is the box counting fractal dimension D_B [69], which is a measure of how densely the fractal fills its image space. It works by covering the signal with the (as many as) necessary N boxes of size S, as S decreases the number of boxes required to cover the signal will increase. If the time series is fractal then the increase will behave as a power law,

$$N \propto S^{-D_B} \tag{8.5}$$

The importance of establishing fractal nature of any time series is that fractals require a degree of organisation. They may have local randomness, but always global statistical determinism revealed by power law scaling properties. Apart from the trivial case where the process is random and governed by the stable laws mentioned above, this places some restrictions on how the underlying process operates, and thus helps us to develop models. Since we have verified that financial time series are not examples of IID stable processes (organised by the CLT), their fractal nature points to some other mechanism for their organisation.

We talk about the scaling properties of fractal and financial time series in more detail in sections 10 and 11 but before this we must investigate another basic related property that has thus far not been considered.

9 Correlation

The second downfall of the random walk model is that price increments are not independent. The Auto-Correlation Function (ACF), $A(t)$, which can be defined as the inverse Fourier transform of the power spectrum [33], is used to measure the structure of the dependence in the price increments:

$$A(t) = \frac{\langle x(t+t')x(t')\rangle - \langle x(t')\rangle \langle x(t+t')\rangle}{\sigma_{x(t')}\sigma_{x(t+t')}} \tag{9.1}$$

$$\approx C \int dk |\widetilde{f}(k)|^2 \exp(-ikt) \tag{9.2}$$

where C is a normalisation constant and $\widetilde{f}$ is the Fourier transform.

The statistics of the power spectrum of a fractal time series can be related to its fractal dimension and also the scaling behaviour of its moments (discussed

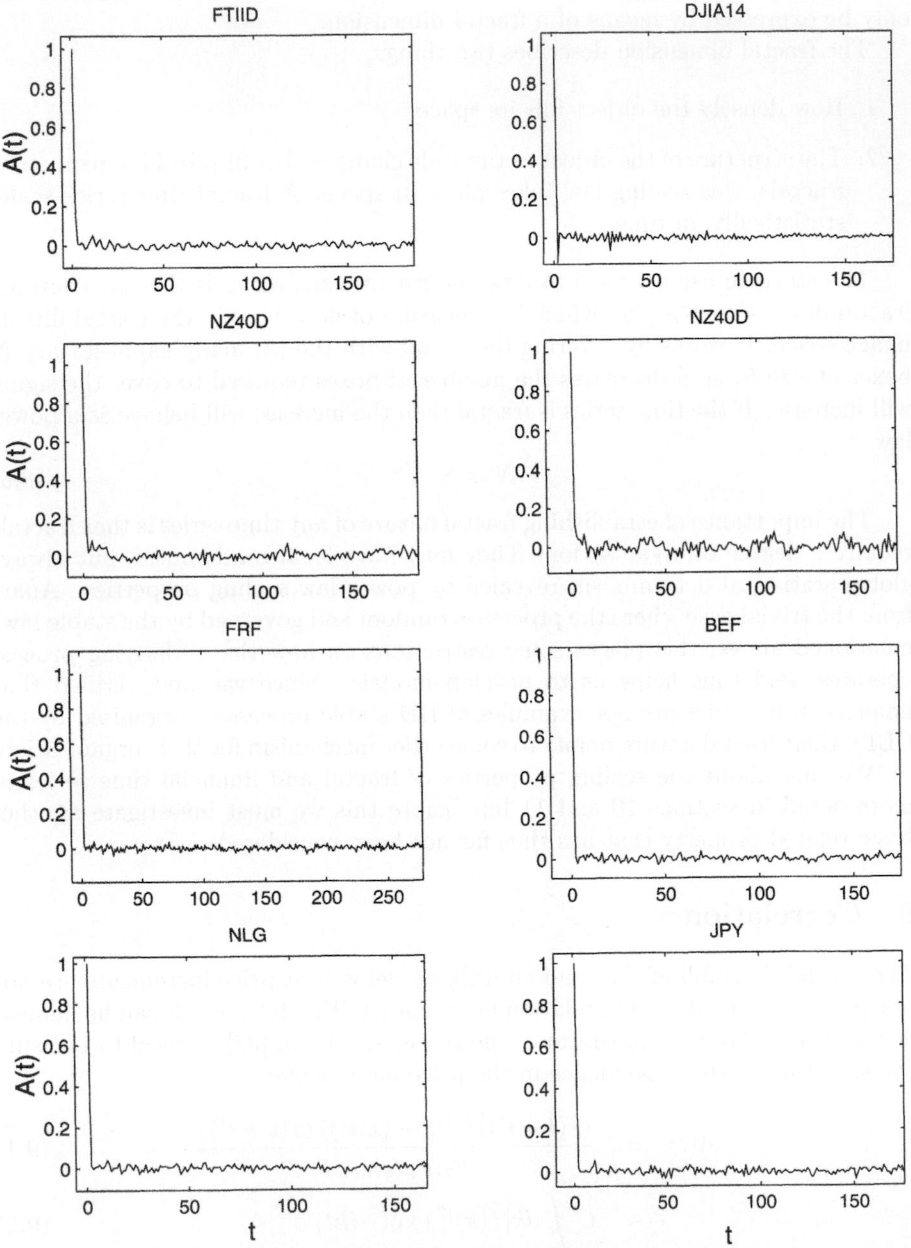

Figure 9. Autocorrelation function of the increments of eight financial time series.

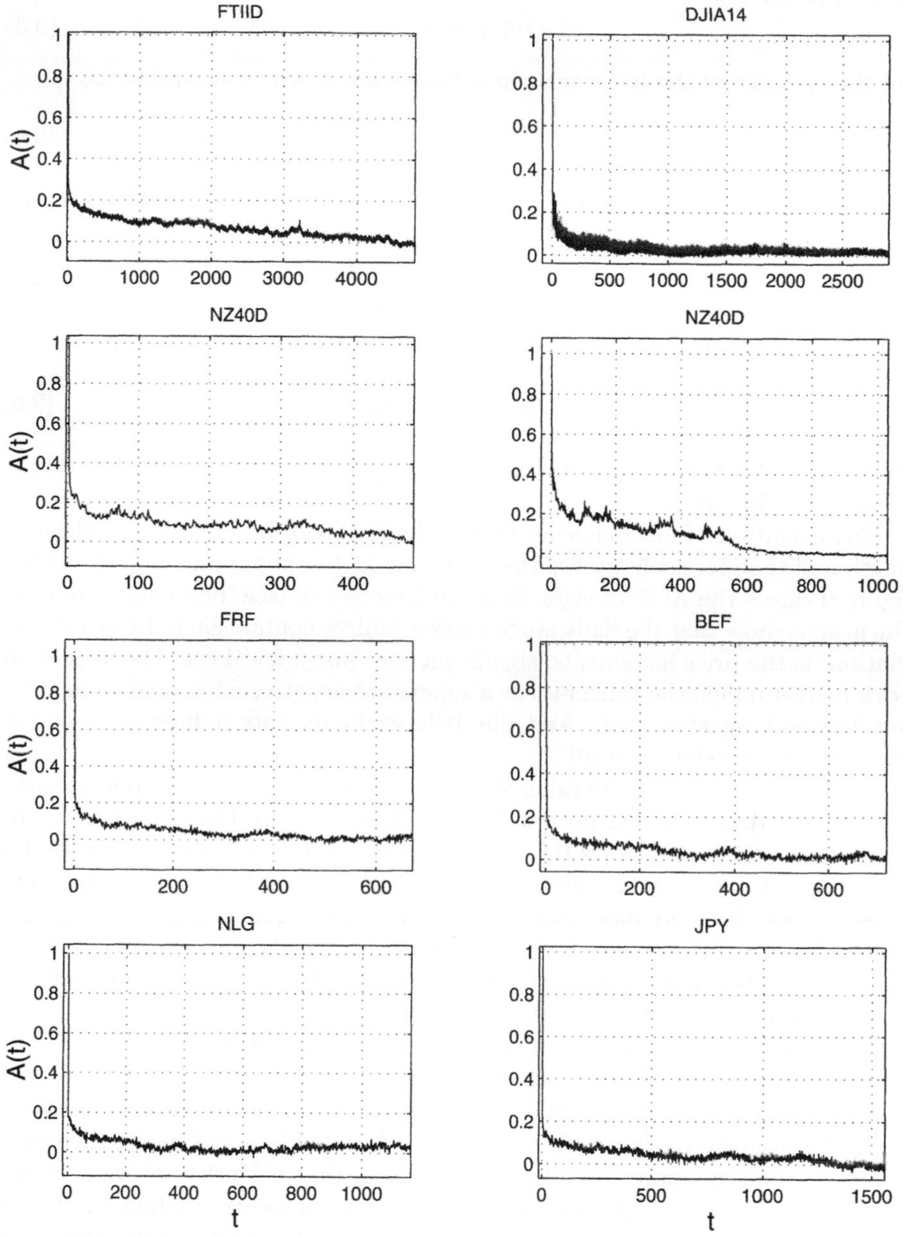

Figure 10. Autocorrelation function of the **absolute** increments of eight financial time series.

again in more detail in sections 10 and 11). If the power spectrum scales according to a power law,

$$|\tilde{f}(k)|^2 \propto k^{-q}, \tag{9.3}$$

and the variance of the increments at a time scale T scales according to,

$$\langle x_T^2 \rangle \sim T^G, \tag{9.4}$$

then the Fourier fractal dimension D_F and G can be directly calculated from q using [25]

$$D_F = 2 - G = \frac{5-q}{2}, \tag{9.5}$$

and the box counting dimension D_B is bound by D_F by [25]

$$D_B < D_F + 1/2. \tag{9.6}$$

Thus the fractal nature of a time series is intimately related to the structure of its time correlations.

The definition of independence is that the autocorrelation of any continuous function of the increments takes the value one at the origin and zero elsewhere. Figure 9 shows the ACF of eight financial time series (see individual captions). The graphs show that the daily stock market indices contain short-lived autocorrelations in the price increments, significant over just a few days. The intra-daily stock market index, the DJIA14, has a significant negative autocorrelation at lag one, but nothing after that. And the daily exchange rate indices show no evidence of autocorrelation at all.

The ACFs of the absolute value of the increments are more consistent amongst the different data sets and decays slowly. There are correlations in the magnitudes, the squares, and hence the volatility of the price increments. This is often referred to by people in finance as volatility clustering. Large increments tend to be followed by more large increments and small increments by small but not necessarily of the same sign. In higher frequency exchange rate data autocorrelations in the raw increments are reported to exist for very short time scales up to approximately 15 minutes [17].

It turns out there is some very short-term correlation in the signs of the increments. For example, for stock market data the probability of a daily price increment having the same sign as the previous one is 0.6 instead of 0.5. These kinds of statistical patterns are investigated thoroughly in [54] using the conditioning entropy measure [74]. It is worth noting that a Markov process defined by the rule that the probability of two particular consecutive values having the same sign is 0.6, has exactly the same ACF structure as the signed increments of the financial data. (See section 15 for details of this experiment.)

Cont [17] shows that the autocorrelations in the square of the price increments can account for the anomalous scaling of the kurtosis (figure 5). If $g(T = N\tau)$ is the autocorrelation of x_τ^2 defined as,

$$g(T) = \frac{\langle\, x_\tau^2(t)x_\tau^2(t+N\tau)\,\rangle - \langle\, x_\tau^2(t)\,\rangle\,\langle\, x_\tau^2(t+N\tau)\,\rangle}{\mathrm{Var}[x_\tau^2]}, \tag{9.7}$$

where τ is some lower time scale to start at (in our case one day) and $\kappa(N\tau)$ is excess kurtosis defined by (8.1) then,

$$\kappa(T) = \frac{\kappa(\tau)}{N} + \frac{6(\kappa(\tau)+2)}{N}\sum_{k=1}^{N}\left(1-\frac{k}{N}\right)g(k). \tag{9.8}$$

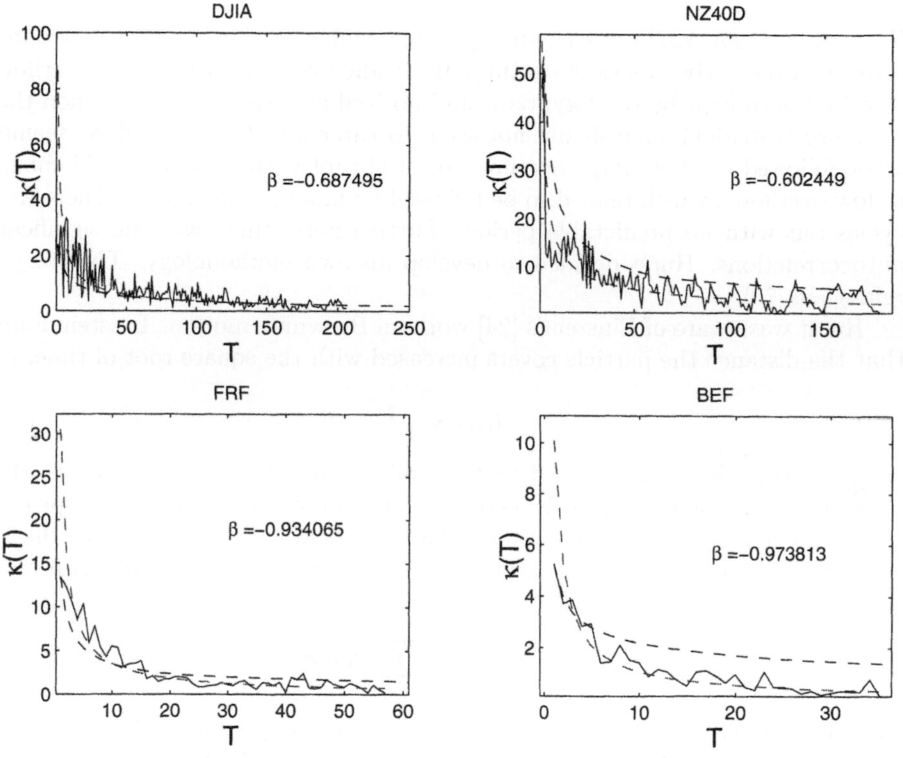

Figure 11. Actual (solid), best fit (light dash) and estimated (dark dash) kurtosis against T.

Figure 11 shows four examples (two stock market and two exchange rate indices) of the empirical scaling of the kurtosis defined by equation (8.1) with best fit and the estimated kurtosis defined by equation (9.8). The fit is not tremendous but close enough to demonstrate that there is a definite link between the correlations in volatility and the slowly decaying kurtosis.

10 Global scaling properties of a 'financial walk'

We now return to the scaling properties of financial time series. In the previous section(s) we examined the leptokurtic nature of the PDF (of price increments) and the correlations structure of the increments and their magnitudes. The analysis carried out thus far is fairly traditional. Next we show that it is possible to explain properties of the time series missed by the previous analysis by looking at the global scaling properties of a financial walk, a type of random walk with signed price increments in place of random numbers. In section 11 we then exploits the ideas further to investigate the phenomenon of *multiscaling*.

10.1 Re-scaled range analysis (RSRA): Measuring memory

H. E. Hurst (1900-1978) was a hydrologist who had worked on the river Nile damn project in the early twentieth century. He studied records of the river overflows that had been kept by the Egyptians and noticed a statistical phenomenon that standard statistical analysis did not seem to cater for. Large overflows seemed to be followed by more large overflows until abruptly, the system would change to low overflows, which tended to be followed by more low overflows. There were cycles but with no predictable period. Furthermore, there were no significant autocorrelations. Hurst decided to develop his own methodology. The story is related in [70].

Hurst was aware of Einstein's [24] work on Brownian motion. Einstein found that the distance the particle covers increased with the square root of time, i.e.

$$R(t) \propto \sqrt{t}, \tag{10.1}$$

where $R(t)$ is the range covered by the walk at time t. For a time series, that is in one dimension, $R(t)$ is defined to be the maximum value of the process minus the minimum up to that time. That is the amount of the real line the 1D random walk has covered. The scaling property (10.1) is just a manifestation of the result,

$$\mathrm{Var}\left(\sum_i x_i\right) = \sum_i \mathrm{Var}(x_i),$$

for uncorrelated (not necessarily independent) x_i. From the ACF of the price increments (figure 9) we should expect that the range of a financial walk scales similarly to the normal random walk.

Hurst's idea was to use this property (equation (10.1)) to test the Nile River's overflows for randomness. Hurst's contribution was to generalise this equation to

$$(R/S)_n = Cn^H, \tag{10.2}$$

where S is the standard deviation for the same n observations and C is a constant.

At this stage we define a Hurst process to be a process that obeys equation (10.2) for some H called the Hurst exponent. The R/S value of equation (10.2) is referred to as the *rescaled range* because it has zero mean and is expressed in terms of local standard deviations.

If the system were independently distributed, then $H = 0.5$. Hurst found that the exponent for the Nile River was $H = 0.91$. The rescaled range was increasing at a faster rate than the square root of time. This meant that the system was covering more distance than a random process would, and therefore the annual discharges of the Nile had to be correlated.

This scaling law (10.2) behaviour is the first connection between Hurst processes and fractal geometry. As previously stated this is important because, apart from the trivial case where $H = 0.5$, this behaviour requires a certain amount of organisation on the part of the underlying process. The source of this organisation is the obvious question and an avenue for research. In the case of the River Nile there is clearly no intelligent mechanism organising the statistics of the overflows. Addressing these type of questions is now a big research area and many researchers are considering interesting new explanations like 'self organised criticality' (SOC)[5].

We have already stated that $H = 0.5$ is consistent with an independently distributed system, but what do other values of H mean? $0.5 < H \leq 1$ implies a *persistent* time series, and a persistent time series is characterised by positive correlations. $0 < H \leq 0.5$ indicates *anti-persistence* which means that the time series covers less ground than a random process. For a system to cover less distance than a random process, it must reverse itself more often than a random process, or be deterministic with slow growth. One can think of a smooth deterministic function as the limit of a stochastic process that reverses itself very quickly but deviates only slightly from its line of best fit. Anti-persistence may be loosely thought of as a mean reverting tendency although this is not formally correct because the system may have no stable mean as such. The Hurst exponent, H, for processes governed by stable (often called fractal or Lévy) distributions (equation 8.2) is related to the dimension, by $H = 1/\mu$ and to the fractal dimension by $H = 2 - D$.

10.1.1 Applying RSRA to financial time series

In this section we apply this analysis to our financial data sets to see if they are, (a) Hurst processes, and (b) random (IID). Peters [70] reports strong evidence that the (daily) Dow Jones Industrial Average financial data set (Jan 2 1888 through Dec 31 1991) is a persistent Hurst process for periods up to four years. A different study [35] concludes that Hurst exponents for financial time series are, in general, not significantly different to 0.5. The test for significance was originally based by Hurst [36] and Feller [28] on the Binomial null hypothesis and the result,

$$(R/S)_N = (N\pi/2)^{1/2}. \tag{10.3}$$

This result has been refined slightly in [70] and a hypothesis test (based on the fact the R/S values are normally distributed random variables) was developed.

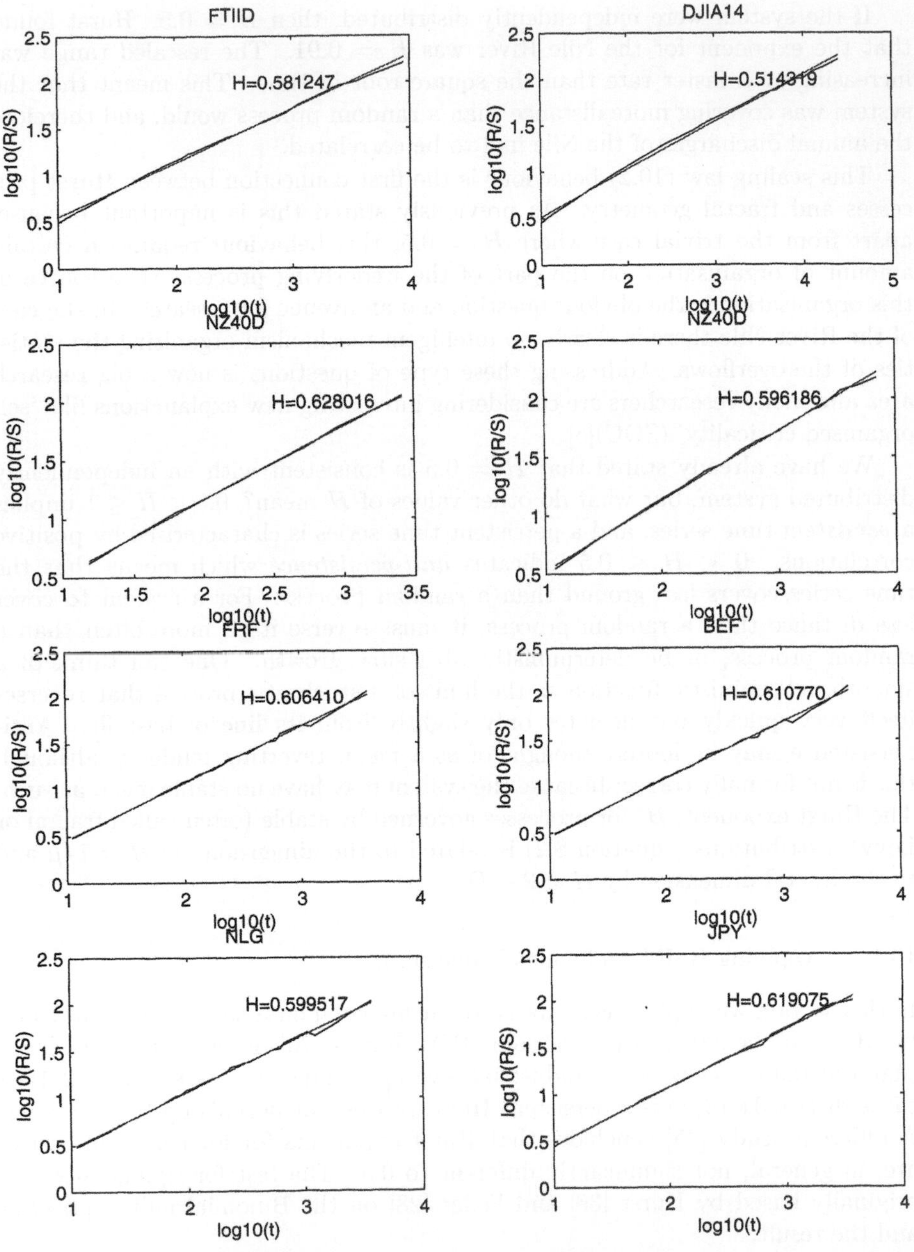

Figure 12. Re-Scaled Range Analysis results for eight financial time series.

The R/S analysis results for the four daily stock market indices and four exchange rate price time series are shown in figure 12. Plotted in the graphs is the log of the rescaled range against the log of the time scale T, thus if equation 10.2 is obeyed then the graph should be a straight line with H as the slope. For the precise methodology used to perform RSRA see 17. The Hurst exponents are in the range $0.51 \leq H \leq 0.64$. The lines fit pretty well and there is no evidence of any kinks in the line. A kink in the line would represent a feature of the data that is nearly periodic: one with a distribution of wavelengths with a spread that is much less than the size of the wavelength. We therefore conclude that our six randomly selected financial processes are Hurst processes.

Figure 13 shows RSRA for the DJIA14 intra-daily data set. One line represents the rescaled range for the increments and the other (the steeper of the two) the rescaled range for the absolute value of the increments. The Hurst exponent for the former is approximately equal to 0.518.

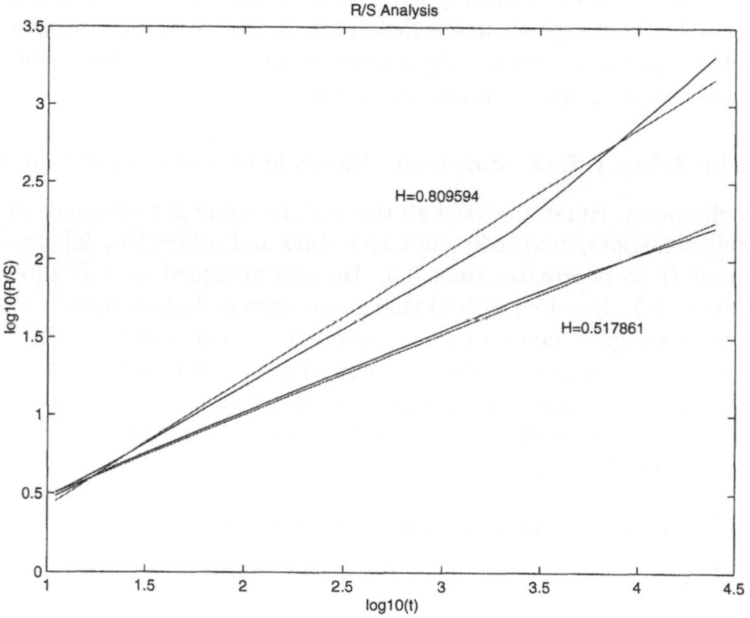

Figure 13. Re-scaled Range Analysis performed on the increments (bottom) and the absolute increments (top) of the DJIA14 intra-daily data set.

Peters [70] also describes RSRA of high frequency 'tick' data and concludes that high frequency data is far more dominated by short-term memory effects and it becomes especially important to take AR(1) residuals prior to performing RSRA, since auto-regressive behaviour can cause a bias in the R/S. This implies that day traders merely react to most recent events. After this is done, markets do not follow random walks even at the 3-minute level, though at this time scale the difference is very small.

At high frequencies we only see pure stochastic processes that resemble white noise, and as we step back and look at lower frequencies, a global structure becomes apparent. The structure is characterised by volatility which behaves similarly to the River Nile overflows: large increments followed by large for a random length of time and then small by small for another random length of time.

10.1.2 Volatility: a Hurst process too?

In the same book [70], Peters describes the same RSRA of (log differences of) a volatility times series and finds that volatility is an anti-persistent process; reversing itself more often than a random process would. This would mean a large increase in volatility has a high probability of being followed by a decrease of unknown magnitude. This is easily verified for the datasets studied above. (See figure 14.) Note that the volatility time series is calculated on non-overlapping time intervals of the original time series. Though the magnitudes of the original increments are persistent (positively autocorrelated), the volatility time series is anti-persistent with negative autocorrelations.

10.2 The Joker Effect: non-periodic cycle explains Hurst processes

After his discovery, Hurst analysed all the data he could get his hands on including rainfall, sun spots, mud sediments, tree rings and others [70, 36]. In all cases, Hurst found H to be greater than 0.5. He was intrigued that H often took a value of about 0.7. Hurst suspected that some universal phenomenon was taking place so to investigate, he carried out some experiments using numbered cards. The values of the cards were chosen to simulate a PDF with finite moments, i.e. $0, \pm 1, \pm 3, \pm 5, \pm 7$ and $\pm$ 9. He first verified that time series generated by summing the shuffled cards gave $H = 0.5$. To simulate a biased random walk, he carried out the following steps.

1. Shuffle the deck and cut it once, noting the number, say n.

2. Replace the card and re-shuffle the deck.

3. Deal out 2 hands of 26 cards, A and B.

4. Replace the lowest n cards of deck B with the highest n cards of deck A, thus biasing deck B to the level n.

5. Place a joker in deck B and Shuffle.

6. Use deck B as time series generator until the joker is cut, then create a new biased hand.

Hurst did 1000 trials of 100 hands and calculated $H = 0.72$ just as he had done in nature. This is an incredible result. Think of the process involved: first the

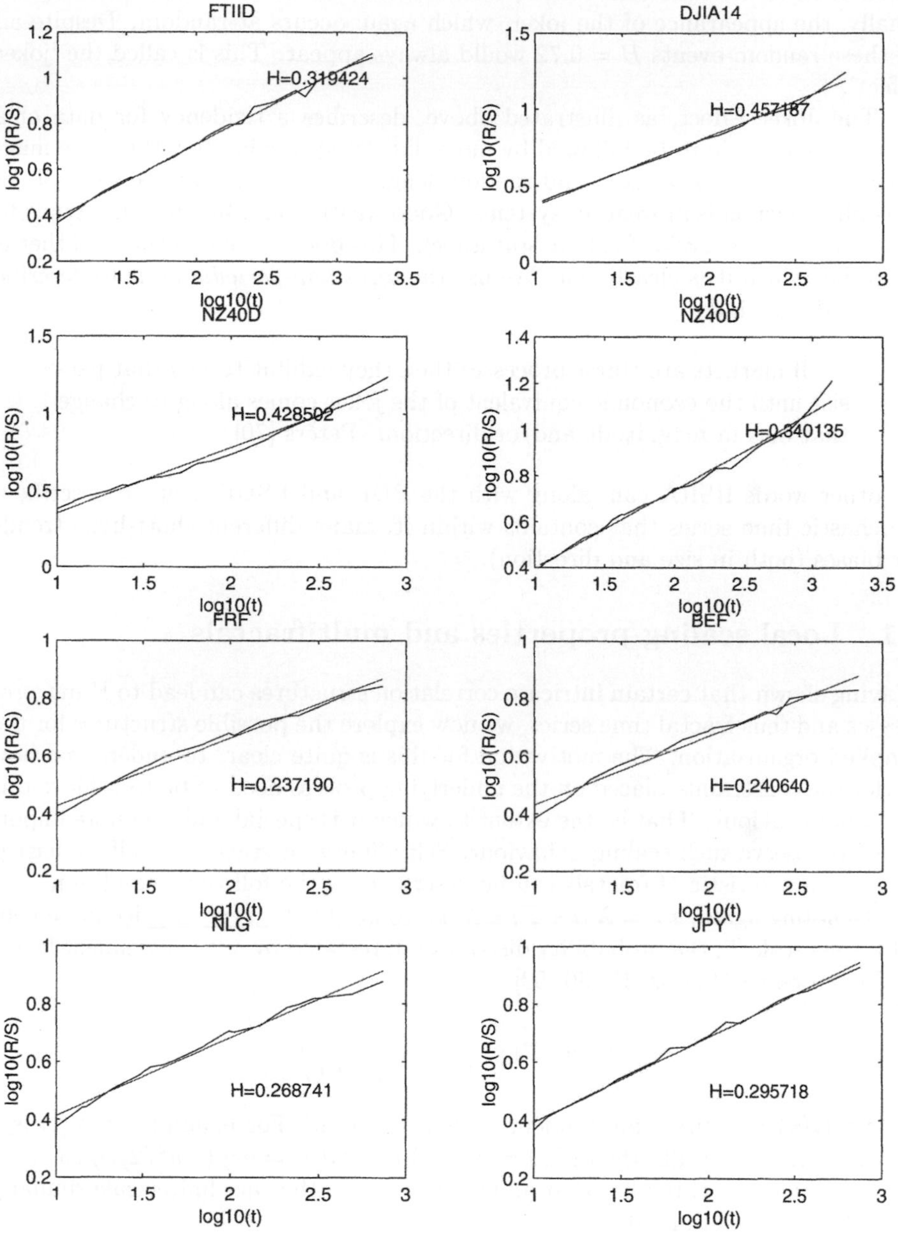

Figure 14. ReScaled Range Analysis results for the **volatility** of eight financial time series.

bias of each hand, which is determined by a random cut of the pack; then, the generation of the time series itself, which is another series of random cuts; and finally, the appearance of the joker, which again occurs at random. Despite all of these random events $H = 0.72$ would always appear. This is called the 'joker effect'.

The Joker Effect, as illustrated above, describes a tendency for data of a certain magnitude to be followed by more data of approximately the same magnitude, but only for a fixed and random length of time. A natural example of this phenomenon is in weather systems. Good weather and bad weather tends to come in waves or cycles (as in a *heat wave*). This does not mean that weather is periodic, which it is clearly not! We use the term *non-periodic cycle* to describe cycles of this kind.

> If markets are Hurst processes then they exhibit trends that persist until the economic equivalent of the joker comes along to change that bias in magnitude and/or direction. -Peters [70]

In other words RSRA can, along with the PDF and PSDF, help to describe a stochastic time series that contains within it, many different short-lived trends or biases (both in size and direction).

11 Local scaling properties and multifractals

Having shown that certain intricate correlation structures can lead to Hurst processes and thus fractal time series, we now explore the possible structures for the implied organisation. The motivation for this is quite clear: to understand how strict the constraints placed on the underlying process need to be to achieve this kind of behaviour. That is, the extent to which it is special and therefore important to observe such scaling behaviour. Self-affinity, or statistical self-similarity, a key characteristic of fractals can be described by the following condition.

Denoting again, $x_T = X(t+T) - X(t)$, to be the de-trended price increment at a time scale T, the probability distribution, rescaled by a lag-dependent factor $\xi(T)$ can be written as [10, 30, 59],

$$P(x_T, T) = \frac{1}{\xi(T)} G\left(\frac{x_T}{\xi(T)}\right), \tag{11.1}$$

where $G(u)$ is a time independent scaling function. For example, if $X(t)$ is a 'normal' random walk, then $\xi(T) = \sigma\sqrt{(T)}$ and $G(u) = \exp(-u^2/2)/\sqrt{2\pi}$.

Equation (11.1) implies that all moments of x_T that are finite scale similarly (for detail see section 16),

$$m_q(T) = \langle |x_T|^q \rangle = A_q \xi(T)^q, \tag{11.2}$$

where A_q is a q-dependent number. Often $\xi(T)$ behaves as a simple power law: $\xi(T) \propto T^{\zeta}$. In this case of a *monofractal* process we have, $m_q \propto T^{\zeta_q}$, with

$\zeta_q \equiv q\zeta$. Here, the parameter ζ is called the Hölder exponent and the function $\zeta(q)$ the scaling function.

Hölder exponents tell us about the regularity of the signal. For $0 < \zeta < 1$, the signal is continuous and for values close to zero the signal is similar to white noise. For $\zeta = 0.5$ the signal is consistent with Brownian motion. For $0.5 < \zeta < 1$, the signal is smoother than Brownian motion. When $\zeta > 1$ the signal breaks up and becomes discontinuous.

A normal random walk (one with a step distribution with a finite second moment), or Wiener Brownian motion (WBM), has $\zeta = 0.5$ at all points and has a scaling function,

$$\zeta(q) = \frac{q}{2}. \tag{11.3}$$

There is another class of signals that possess this monofractal property, known as fractional Brownian motions (FBM) and their discrete analogue, fractional Gaussian noises (FGN) inspired by Mandelbrot and Ness [60]. FBMs are Gaussian fractionally integrated processes, that behave as,

$$P(x_T, T) = \frac{1}{T^H} G\left(\frac{x_T}{T^H}\right), \tag{11.4}$$

defined by the convolution of Gaussian noise $\eta(t)$ and some power law function of t:

$$B_H(t) = \int_0^t (t-s)^{H-\frac{1}{2}} \eta(s) ds, \qquad 0 \leq H \leq 1, \tag{11.5}$$

where H denotes again the Hurst exponent. This is equivalent to the following definition via the Fourier transform, denoted by $\tilde{B}$.

$$\widetilde{B}_H(\omega) = (i\omega)^{-0.5-H} \widetilde{\eta}(\omega), \qquad 0 \leq H \leq 1, \tag{11.6}$$

which is also a widely accepted definition of fractional differ-integration [66, 73, 42]. It is useful to know that this is also the same as filtering the power spectrum of η:

$$||B_H(t)|| = \widetilde{B}\widetilde{B}^* = \frac{1}{|\omega|^{1+2H}}. \tag{11.7}$$

From this equation one can see that the smoothing works by filtering out the power of the signal at high frequencies. The correlation between two equal and non-overlapping increments of FBM is [60, 22],

$$C(t) = \langle (B_H(0) - B_H(-t)) (B_H(t) - B_H(0)) \rangle / B_H(t)^2 = 2^{2H-1} - 1, \tag{11.8}$$

which is (notably) independent of t, illustrating the self-affine nature of the process. The autocorrelation at a lag t is given by [60],

$$C(t) = \frac{1}{2}\left(|t+1|^{2H} - 2|t|^{2H} - |t-1|^{2H}\right), \tag{11.9}$$

which is clearly zero for $H = 1/2$ and for $H \neq 1/2$ can be approximated by the power law,

$$C(t) \sim t^{2H-2}. \tag{11.10}$$

Therefore FBM has a scaling function,

$$\zeta(q) = \zeta q = Hq. \tag{11.11}$$

11.1 Multiscaling

When the function $\xi(q)$ behaves as (11.2) with $\zeta_q \neq \zeta q$ then the process x_T is said to be multiscaling [30, 56, 59, 10, 68, 38, 3]. One of the implications of this is that there is, instead of one Hölder exponent (ζ) to describe the whole process, now an intricate mixture of local (coarse) Hölder exponents.

We can expose mixtures of Hölder exponents by examining multiscaling properties because different values of q put an emphasise on different sized x values. That is, when we examine the scaling of $|x_T|^q$ for large q, only the large x values are contributing. The extent to which this happens is related to the regularity of the signal. Smooth signals have small increments that are of similar magnitude. Jagged signals that go up and down more violently have a greater variety of magnitudes and therefore a greater capacity for the larger ones to dominate.

What is the scaling function of financial data like? Figure 15 shows the partition function, defined as:

$$S_q(T) \propto T\, m_q(T) = \sum |x_T|^q \propto T^{\zeta(q)+1}, \tag{11.12}$$

for values of q ranging from 1.5 to 4 in steps of 0.5. The dashed lines represent the theoretical scaling slopes for WBM as described by equation (11.3). Since the lines are straight, the moments scale as power laws, indicating that the data are self-affine. The scaling function is not the same as that of the random walk or WBM and more importantly, it is not linear in q. In the case of the DJIA14, the higher frequency index, there is a clear *crossover* period caused by the strong negative autocorrelations in the higher frequency (half hourly) data.

We therefore conclude that our financial data are indeed multifractal. Some recent papers [30, 56, 17, 19, 4] suggest that this is a fruitful avenue of research in the sense that a lot of organisation is required for this property to hold: the distribution of Hölder exponents for instance.

The scaling function $\zeta(q)$, determined by the moments, can be used to estimate the *multifractal spectrum* which tells us the distribution of Hölder exponents [30]. The multifractal spectrum, $f(\alpha)$, and the scaling function, ζ_q, are related through the Legendre transform [30, 56, 59],

$$\hat{f}(\alpha) = \min_{\alpha} [q\alpha - \hat{\lambda}(q)]. \tag{11.13}$$

Examples of systems known to exhibit multiscaling are turbulence [32], Internet traffic [27], gestural expressionist art [65] and many others [57]. Whereas

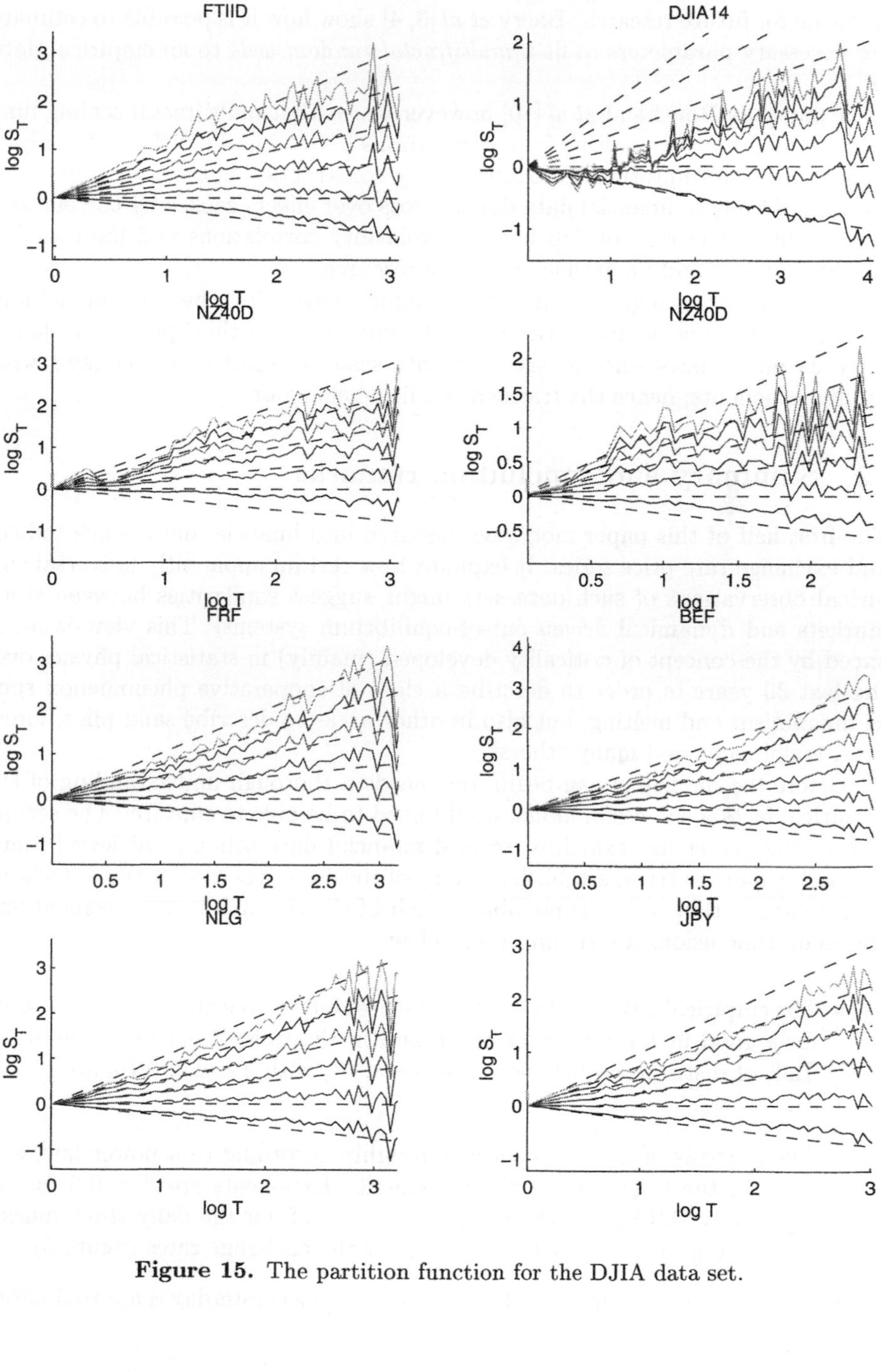

Figure 15. The partition function for the DJIA data set.

there is strong theoretical evidence for multiscaling in turbulence [32], there is as yet no theory connecting the phenomenon with financial time series. This is a direction for future research. Bacry *et al* [3, 4] show how it is possible to estimate the necessary parameters to fit a *multifractal random walk* to an empirical data set.

A paper by Bouchaud *et al* [10] however warns that multifractal scaling may be a trivial property resulting from correlations in the volatility. They describe a model that is asymptotically monofractal by construction but that exhibits multiscaling similar to financial data due to cross over effects caused by correlations in volatility. The relationship between volatility correlations and multiscaling remains unclear and an avenue for future research.

Bouchaud *et al* suggest that the cumulants better describe their model and perhaps also financial time series since the cumulants of their process scale exactly as power laws whereas the moments scale as sums of power laws with similar exponents; hence the transient scaling behaviour.

12 Summary and concluding remarks

The first half of this paper motivates research into financial data; stock indices and exchange rate price series. It explains how certain apparently universal empirical observations of such data sets might suggest similarities between stock markets and dynamical driven out-of-equilibrium systems. This view is influenced by the concept of criticality developed (mainly) in statistical physics over the last 30 years in order to describe a class of cooperative phenomenon such as magnetism and melting, but also in other areas to describe sand piles, forest fires, traffic jams and many others.

Before the modelling can begin, one needs a thorough understanding of the empirical facts a successful model would need to be able to capture. The second half of this paper has examined several financial data sets and reviewed many research papers to try to establish as many of the empirical regularities associated with financial time series as possible. Much of the research is in agreement and the main conclusions are summarised below.

- The empirical PDF has Lévy style features but exponents are not consistent for a good fit both near the mean and in the tails; it is one or the other. Tails of distribution behave as, $p(x) \sim x^{-\alpha}$, with $3 < \alpha < 4$ (figures 4 and 6).
- The kurtosis of increments scales roughly according to a power law, $\kappa \sim T^{-\beta}$, in the time scale of the increment. Exponents are $\beta \approx 0.3$ for the intra-daily DJIA14, in the range $0.5 < \beta < 0.6$ for the daily stock market indices and in the range $0.9 < \beta < 1$ for the exchange rates (figure 5).
- The probability of moving the same direction as yesterday is approximately 0.6 instead of 0.5.

- The autocorrelation function of the signed increments is not significantly different to zero for time-scales above half and hour. (figure 9).
- The autocorrelation function of the absolute increments decays slowly, reaching zero typically at around 500 days (2 years) (figure 10).

There is additionally evidence of:

- Slight persistence in increments, though not statistically significant (figure 12).
- Strong anti-persistence in the volatility increments (figure 14).
- Multiscaling (figure 15).

Several areas seem obvious main targets for future research. The first is the statistical issue surround the phenomenon of multiscaling and how it is affected by the volatility clustering. The second and perhaps most important area is the need for more simple 'toy' models like those described in [6, 18, 40, 50, 31, 39] to understand how the concept of criticality and universality of markets is made possible. This will be the subject of a forthcoming paper.

Appendix

13 The Central Limit Theorem (CLT)

Consider a number N of independent and identically distributed (IID) random variables x_n with $n = 1, ..., N$, with probability density function, $p(x)$. Let,

$$X_N = \sum_{n=1}^{N} x_n. \tag{13.1}$$

Given that the first two moments, $\langle x \rangle$ and $\langle x^2 \rangle$ are finite, the mean and variance of the X depend linearly on N,

$$\bar{X} = \langle x \rangle N, \quad \bar{X^2} - \bar{X}^2 = (\langle x^2 \rangle - \langle x \rangle^2)N. \tag{13.2}$$

Up to a translation of reference we may suppose that $\langle x \rangle = 0$. X_N behaves as $N^{1/2}$ and it is the distribution of the variable, $X_N/N^{1/2}$ that admits a limiting form. The distribution of $X/N^{1/2}, f(X)$, given by the joint distribution for the summands is,

$$f(X) = \int ... \int \prod_{n=1}^{N} p(x_n)dx_n \, \delta\left(\frac{X_N}{\sqrt{N}} - N^{1/2}\sum_{n=1}^{N} x_n\right). \tag{13.3}$$

Using an integral representation of the δ-function,

$$\delta(x) = \frac{1}{2\pi}\int_{k=-\infty}^{+\infty} \exp(-ikx)dk, \tag{13.4}$$

equation (13.3) may be written as,

$$f(X) = \frac{1}{2\pi}\int_{k=-\infty}^{+\infty} dk \exp(ikX/\sqrt{N}) \left(\int dx p(x) \exp(-ikx/\sqrt{N})\right)^N . \quad (13.5)$$

This expression involves the Fourier transform (characteristic function) $\tilde{p}(k)$ of $p(x)$. For large N only the behaviour of $\tilde{p}(k)$ close to $k = 0$ is important, since $|\tilde{p}(k)| \leq 1$.

$$\tilde{p}(k/\sqrt{N})^N = [1 - \frac{1}{2}\langle x^2\rangle k^2/N + \mathrm{O}(N^{3/2})]^N \rightarrow \exp(-\langle x^2\rangle k^2/2) \quad (13.6)$$

The Gaussian PDF is then recovered by integrating 13.5 over k.

14 De-trending process for financial time series

Here we justify using the auto-regressive one log difference de-trending process (section 7) for all the financial time series even though the exchange rates do not show the same exponential trends as the stock markets. The are several reasons that this is a good idea. The first is for consistency: consistent treatment of all data sets and consistency with most other published quantitative research. Another reason is a general argument often used by physicists for signal processing. The details follow below but the general argument goes something like this.

Most systems are translationally time invariant and as such can be written so that its solutions are the eigenvectors of a linear operator that commutes with a time translation operator. Autonomous systems have solutions in the form of exponentials (including complex). Also, any two linear operators that commute share eigenvectors, so in this way any operator that commutes with the time translation operator is autonomous and has exponential eigenvectors; therefore the corresponding system has exponential solutions.

If A is a linear operator such that,

$$A\underline{\mathrm{x}} = \lambda\underline{\mathrm{x}}, \quad (14.1)$$

then $\underline{\mathrm{x}}$ is an eigenvector of A and λ is an eigenvalue. Suppose a system is governed by a equation like,

$$Af(t) = cf(t), \quad (14.2)$$

where c is a constant. For example, the differential equation,

$$\frac{d^2 f}{dt^2} + \omega^2 f(t) = 3f(t), \quad (14.3)$$

or a stochastic differential equation, then solutions to the system are eigenvalues of A. A system is translationally invariant (autonomous), when if $f(t)$ is a solution, $f(t-\alpha)$ is also a solution. For example,

$$\frac{df}{dt} = 3, \quad (14.4)$$

is autonomous but,

$$\frac{df}{dt} = t, \tag{14.5}$$

isn't.

14.1 Statement

$Af = cf$ is autonomous if and only if (iff) A commutes with T_α, i.e.,

$$[A, T_\alpha] = AT_\alpha - T_\alpha A = 0, \tag{14.6}$$

where T_α is the time translator operator,

$$T_\alpha f(t) = f(t - \alpha). \tag{14.7}$$

14.2 Proof

We must first prove that if $f(t)$ is the solution of the system $Af = cf$ where A commutes with the time translation operator T_α, then $f(t-\alpha)$ is also a solution. Say $f(t)$ solves $Af = cf$, and $[A, T_\alpha] = 0$, then, $T_\alpha Af = cT_\alpha f$, and $A(T_\alpha f) = c(T_\alpha f)$, but, $T_\alpha f = f(t-\alpha)$, so $Af(t-\alpha) = cf(t-\alpha)$, i.e. $f(t-\alpha)$ is a solution.

We must also prove that whenever $f(t)$ and $f(t-\alpha)$ are solutions, $[T_\alpha, A] = 0$. If $Af(t) = cf(t)$ and $Af(t-\alpha) = cf(t-\alpha)$ then the latter can be written in terms if the time translator operator as, $A(T_\alpha f(t)) = c(T_\alpha f(t)) = T_\alpha(cf(t)) = T_\alpha Af(t)$. Thus, $AT_\alpha = T_\alpha A = 0$.

14.3 Theorem

If $[A, B] = 0$ then every eigenvector of A is an eigenvector of B.

14.3.1 Proof

If $\underline{x}$ is an eigenvector of A then,

$$A\underline{x} = c\underline{x} \tag{14.8}$$

$$BA\underline{x} = cB\underline{x} \tag{14.9}$$

$$AB\underline{x} = cB\underline{x}. \tag{14.10}$$

If eigenvectors are non-degenerate (each eigenvector has unique eigenvalue) then we say, $B\underline{x}$ is an eigenvector of with eigenvalue c, so $B\underline{x}$ is a multiple of $\underline{x}$, so $\underline{x}$ is an eigenvector of B.

15 Autocorrelation function of a Markov process

A Markov process defined by the rule: the probability that the price movement at time $t+1$ has the same sign as the price movement at time t is equal to p has the following properties.

Denote r to be the random variable that is the length of a sequence of consecutive movements in the same direction, then the mean of the process is, $\langle r \rangle = p/(1-p)$ and the standard deviation is, $\sigma_r = p/(1-p)^2$. Empirically the mean values of r are in the range, $1.35 > \langle r \rangle > 1.65$ indicating that the p value is in the range, $.0.574 > p > 0.623$. The standard deviations are around 3.75 which is larger than $\langle r \rangle$ and for this reason we should not expect any near period cycles.

Figure 16 shows that the correlation structure of the de-trended increments of the FTIID index is similar to that of a Markov process defined with the same p value ($p = 0.5836$). The autocorrelation structure depends also of course on the shape of the conditional distributions, which in this experiment were taken to be uniform. The point of this section in the appendix was just to show that this feature (the Markov feature) can capture the correlation structure of the daily financial time series.

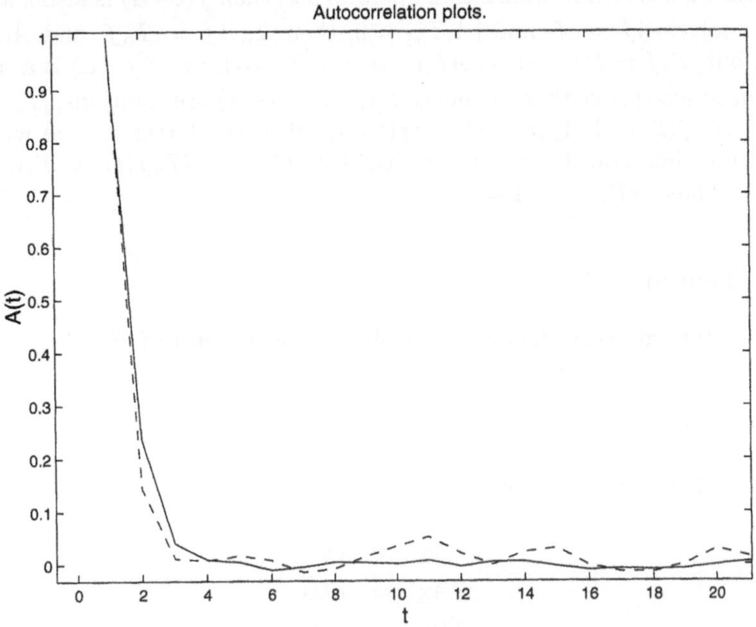

Figure 16. Autocorrelation of FTIID (dashed) versus synthetic Markov process (solid).

16 Proof of equation (11.2)

From equation (11.1) the moments m_q are defined as,

$$\begin{aligned} m_q = \langle |x_T|^q \rangle &= \int |x_T|^q \frac{1}{\xi(T)} G\left(\frac{x_T}{\xi(T)}\right) dx_T && (16.1) \\ &= \int |y\xi(T)|^q G(y) \frac{1}{\xi(T)} \xi(T) dy && (16.2) \\ &= |\xi(T)|^q \int |y|^q G(y) dy && (16.3) \\ &= |\xi(T)|^q A_q && (16.4) \end{aligned}$$

17 R/S analysis applied

The following is a step by step methodology for applying R/S analysis to stock market data. Note that the AR(1) [70] notation used below stands for auto regressive process with 1-daily linear dependence. Thus taking AR(1) residuals of a signal is equivalent to plotting the signals 1 day out of phase and taking the day to day linear dependence out of the data.

1. Prepare the Data. Take AR(1) residuals of log ratios. The log ratios account for the fact that price changes are relative, i.e. depend on price. The AR(1) residuals remove any linear dependence, serial correlation, or short-term memory which can bias the analysis.

$$\begin{aligned} V_t &= \log(P_t/P_{t-1}) && (17.1) \\ X_t &= V_t - (c + mV_{t-1}) && (17.2) \end{aligned}$$

   The AR(1) residuals are taken to eliminate any linear dependency.

2. Divide this time series (of length N) up into A sub-periods, such that the first and last value of time series are included i.e. $A \times n = N$. Label each sub-period I_a with $a = 1, 2, 3, ..., A$. Label each element in I_a with $N_{k,a}$ where $k = 1, 2, 3, ..., n$. For each I of length n, calculate the mean

$$e_a = (1/n) \sum_{k=1}^{n} N_{k,a} \quad (17.3)$$

3. Calculate the time series of accumulated departures from the mean for each sub interval.

$$Y_{k,a} = \sum_{i=1}^{k} (N_{i,a} - e_a) \quad (17.4)$$

4. Define the range as

$$R_{I_a} = \max(Y_{k,a}) - \min(Y_{k,a}) \tag{17.5}$$

where $1 \leq k \leq n$.

5. Define the sample standard deviation for each sub-period as

$$S_{I_a} = \sqrt{\frac{1}{n}\sum_{k=1}^{n}(N_{k,a} - e_a)^2} \tag{17.6}$$

6. Each range, R_{I_a} is now normalised by dividing by its corresponding S_{I_a}. Therefore the re-scaled range for each I_a is equal to R_{I_a}/S_{I_a} From step 2 above, we have A contiguous sub-periods of length n. Therefore the average R/S value for each length n is defined as

$$(R/S)_n = \frac{1}{A}\sum_{a=1}^{A}(R_{I_a}/S_{I_a}) \tag{17.7}$$

7. The length n is then increased until there are only two sub-periods, i.e. $n = N/2$. We then perform a least squares regression on $\log(n)$ as the independent variable and $\log(R/S)$ as the dependent variable. The slope of the equation is the estimate of the Hurst exponent, H.

Bibliography

1. The data were purchased by De Montfort University from Global Financial data (http://www.globalfindata.com/) in 1999.

2. Louis Bachelier. Theory of speculation. In P. Cootner, editor, *The Random Character of Stock Market Indices.* Cambridge, MA: M.I.T. Press, 1964. Originally Bachelier's PhD thesis completed in 1900.

3. E. Bacry. A multifractal random walk. submitted to IEEE Trans. on Inf. Theory.

4. E. Bacry, J. Delour, and J.F. Muzy. Modelling financial time series using multifractal random walks. To be published in Physica A (proceedings of the Nato Advanced Research Workshop on 'Application of Physics in Economic Modelling', Prague, 2001.

5. Per Bak. *How nature works.* Copernicus/ Springer-Verlag, 1996.

6. Per Bak, M. Paczuski, and M. Shubik. Price variations in a stock market with many agents. cond-mat/9609144, Sept 1996.

7. G. L. Baker and J. P. Gollub. *Chaotic dynamics - an introduction.* Cambridge University Press, 1990.

8. R. Barro, E. Fama, D. Fischel, A. Meltzer, R. Roll, and L. Telser. Black monday and the future of financial markets, 1989. edited by R.W. Kamphuis, Jr., R.C. Kormendi and J.W.H. Watson (Mid American Institute for Public Policy Research, Inc. and Dow Jones-Irwin, Inc., 1989).

9. F. Black and M. Scholes. The pricing of options and corperate liabilities. *Journal of Political Economy,* May/June 1973.

10. Jean Bouchaud, Marc Potters, and Martin Meyer. Apparent multifractality in financial time series. *Cond-mat,* June 1999.

11. Jean-Philippe Bouchaud. Power laws in economy and finance: some ideas from physics. *Proceedings of the 2000 Santa Fe conference* (to be published) *in Quantitative Finance,* 2000.

12. Jean-Philippe Bouchaud and Antoine Georges. Anomalous diffusion in disordered media: Statistical mechanisms models and physical applications. *physics Reports (Review section of Physics Letters),* 195(4-5):127–293, 1990.

13. Robert Brown. A brief account of microscopical observations. Unpublished, 1827.

14. Mark Buchanan. *Ubiquity.* Weidenfield and Nicolson, 2000.

15. Philippe Carmona and Laure Coutin. Fractional brownian motion and the markov property. *Electronic communications in probability,* 3:95–107, 1998.

16. A. Compte. Stochastic foundations of fractional dynamics. *Physics review E,* 1996.

17. Rama Cont. Scaling and correlation in financial data. *European physical journal B,* 1997. cond-mat/9705075.

18. Rama Cont and Jean-Philippe Bouchaud. Herd behaviour and aggregate fluctuations in financial markets. *Journal of Macro-economic Dynamics,* 4(2):2000, 2000.

19. Rama Cont, Marc Potters, and Jean-Philippe Bouchaud. *Scale invariance and beyond,* chapter Scaling in stock market data: stable laws and beyond. Springer, 1997.

20. Theodore A. Cook. *The Curves of Life.* Dover books, 1979. A Dover reprint of a classic 1914 book.

21. B. Davidovich and I. Procaccia. Conformal theory of the dimensions of diffusion limited aggregates. *Europhysics Letters,* 48:547, 1999. # chao-dyn/9812026.

22. P. Devynck, G. Wang, and G. Bonhomme. The hurst exponent and long-time correlation. In *27th EPS Conference on Contr. Fusion and Plasma Phys., Budapest, 12-16 June 2000*, volume 24B, pages 632–635. ECA, 2000.

23. Stephane Douady and Yves Couder. *Phyllotaxis as a self-organised growth process*, pages 341–352. Plenum press, 1993.

24. A. Einstein. The theory of the Brownian movement. *Ann. der Physik*, page 17:549, 1905.

25. A. K. Evans. The Fourier dimension and the fractal dimensions. *Chaos, Solitons and Fractals*, 9(12), 1998.

26. J. Doyne Farmer. Physicists attempt to scale the ivory towers of finance. *Computing in science and engineering*, pages 26–39, Nov/Dec 1999.

27. A. Feldmann, A. Gilbert, and W. Willinger. Data networks as cascades: Investigating the multifractal nature of internet wan traffic, 1998. In Proc. of the ACM/SIGCOMM'98, pages 25–38, 1998.

28. W. Feller. *An Introduction to Probability and its Applications*, volume 1. John Wiley & Sons, New York, 3rd edition, 1968.

29. Richard P Feynman. *The Meaning of it All.* Penguin, 1999.

30. Fisher, Calvet, and Mandelbrot. A multifractal model of asset returns. *Journal of finance*, 1998.

31. Henrik Flyvbjerg. Mean field theory for a simple model of evolution. *Physical review letters*, 71(24):4087–4090, December 1993.

32. U. Frish. *The Legacy of Kolmogorov.* Cambridge U.P., Cambridge, UK, 1995.

33. Daniel T. Gillespie. The mathematics of brownian motion and johnson noise. *American association of physics teachers*, 1996.

34. Hari M. Gupta and José R. Campanha. The gradually truncated lévy flight for systems with power law distributions. *Physica A*, 268:231–239, 1999.

35. John S. Howe, Deryl W. Martin, and Bob G. Wood. Much ado about nothing: Long term memory in pacific rim equity markets. *International Review of Financial Analysis*, 8(2):139–151, 1999.

36. H. Hurst. Long term storage capacity of reservoirs. *Trans. Amer. Soc. Civil Eng.*, pages 770–808, 1950.

37. Jessica James. Modelling the money markets. *Physics World*, September 1999.

38. W. Jeżewski. Multiscaling and multifractality in an one-dimensional ising model. *Euro. Phys. J. B*, 19:133–138, 2001.

39. A. Johansen, O. Ledoit, and D. Sornette. Crashes as critical points. *International Journal of Theoretical and Applied Finance in press (1999); preprint http://xxx.lanl.gov/abs/cond-mat/9810071.*, 1999.

40. Andres Johansen and Didier Sornette. A hierarchical model of financial crashes. *Physica A*, 261(3-4):581–598, Aug 1998.

41. M. Kendall. The analysis of economic time-series-part i: Prices. *Journal Statistical Society A, mathematical Statistics*, 116:11–25, 1953.

42. V. Kiryakova. *Generalized fractional calculus.* Longman, 1994.

43. Ron Knott. Fibonacci numbers and nature. Web site of the University of Surrey. http://www.ee.surrey.ac.uk/Personal/R.Knott/Fibonacci/-fibnat.html.

44. Ismo Koponen. Analytic approach to the problem of convergence of truncated Lévy flights towards the Gaussian stochastic process. *Physical review E*, 52(1), July 1995.

45. I. Koutrouvelis. Regression-type estimation of the parameters of stable laws. *Journal of American Statistical Association*, 75:918–928, 1980.

46. I. Koutrouvelis. An iterative procedure for the estimation of the parameters of stable laws. *Communications in Statistics - Simulation and Computation*, B10:17–28, 1981.

47. J. Kurths, A. Voss, P. Saparin, A. Witt, H. Kleiner, and N. Wessel. Quantitative analysis of heart rate variability. *Chaos*, 5:88, 1995.

48. L. Laloux, M. Potters, R. Cont, J. Aguilar, and J. Bouchaud. Are financial crashes predictable? *Europhysics Letters*, January 1999.

49. L.D. Landau and E.M. Lifshitz. *Fluid Mechanics*, volume 6. Butterworth-Heinemann, 1987.

50. Moshe Levy and Sorin Solomon. Dynamical explanation for the emergence of power law in a stock market model. *International Journal of Modern Physics C*, 7(1):65,72, 1996.

51. Paul Lévy. *Théorie de l'addition des variables aléatoires.* Paris: Gauthier-Villars, 1937-54. Second edition, 1954.

52. M. J. Lighthill. *Fourier Analysis and Generalised Functions.* Cambridge University Press, 1962.

53. Alexander Lipton-Lifschitz. Predictability and unpredictability in financial markets. *Physica D*, 133:321–347, 1999.

54. M. D. London, A. K. Evans, and M. J. Turner. Conditional entropy and randomness in financial time series. *Quantitative Finance*, 1(4):414–426, June 2001.

55. E. Lorenz. Deterministic non-periodic flow. *J. Atmospheric Sci.*, 20:130–141, 1963.

56. Mandelbrot, Calvet, and Fisher. Multifractality of the dem/$ exchange rate. *Scientific American*, 1999.

57. B. Mandelbrot. *Multifractals and 1=f Noise.* Springer-Verlag, New York, 1999.

58. Benoit B Mandelbrot. *The fractal geometry of nature.* W. H. Freeman and Co., New York, 1982.

59. Benoit B. Mandelbrot. *Fractals and scaling in finance.* Springer, 1997.

60. Benoit B. Mandelbrot and John W. Van Ness. Fractional brownian motions, fractional noises and applications. *SIAM Review*, 10(4):423–437, Oct 1968.

61. Rosario N. Mantegna and H. Eugene Stanley. Stochastic process with ultra slow convergence to gaussian: the truncated lévy flight. *Physical review letters*, 73(22):2946–2949, November 1994.

62. P. Masani. *Norbert Wiener: Collected Works*, volume 1. MIT Press (Cambridge, MA), 1976.

63. R. May. Simple mathematical models with very complicated dynamics. *Nature*, 261:459–467, 1976.

64. J. McCulloch. Simple consistent estimators of stable distribution parameters. *Communications in Statistics - Computation and Simulation*, 15:1109–1136, 1986.

65. J. R. Mureika, Gerald C. Cupchik, and Charles C. Dyer. Looking for fractals in gestural expressionist paintings. In *Proceedings of the XVI Congress of the International Association of Empirical Aesthetics.* New York, August 2000.

66. Nigmatullin. Fractional integral and it's physics interpretation. *Theoretical and mathematical physics*, 1992.

67. L. Oddershede, P. Dimon, and J. Bohr. Self-organised. *Physical Review Letters*, 71:3107–31110, 1993.

68. Michele Pasquini and Maurizio Serva. Multi-scale behaviour of volatility autocorrelations in a financial market. *cond-mat/9810232*, Oct 1998.

69. Heinz-Otto Peitgen, Hartmut jürgens, and Dietmar Saupe. *Chaos and fractals: new frontiers of science.* New York;London:Springer-Verlag, 1992.

70. E. Peters. *Fractal market analysis.* Wiley, 1994.

71. Tony Plummer. *Forecasting Financial Markets.* Kogan Page, 1990.

72. Marc Potters, Rama Cont, and Jean-Philippe Bouchaud. Financial markets as adaptive ecosystems. *Europhysics letters*, 41(3):239–244, Feb 1998.

73. R. S. Rutman. On the physical interpretation of fractional integration. *Theoretical and mathematical physics*, 1995.

74. C. E. Shannon and W. Weaver. *The mathematical theory of communication.* The university of Illinois press, 1949.

75. Didier Sornette, Anders Johansen, and Jean-Philippe Bouchaud. Stock market crashes, precursors and replicas. *Journal of Physics I*, 6(1):167–175, Jan 1996.

76. Ian Stewart. *Does God Play Dice.* Penguin, 1997.

77. Alan Stuart and Keith Ord. *Distribution Theory*, volume 1. Edward Arnold, sixth edition, 1994.

78. Constabtino Tsallis. Lévy distributions. *Physical Review*, 1997.

79. Nelson Wax, editor. *Selected Papers on Noise and Stochastic Processes.* Dover, 1954.

80. Paul Wilmott, Sam Howison, and Jeff Dewynne. *The mathematics of financial derivatives.* Cambridge, 1995.

81. H. Working. Prices of cash wheat and futures since 1883. *Wheat Studies*, 2(3), 1934.

Analysis of the Limitations of Fractal Dimension Texture Segmentation for Image Characterisation

Martin J. Turner and Jonathan M. Blackledge

ISS, SERC, Hawthorn Building, De Montfort University, Leicester LE1 9BH

Abstract

An overview of the Fourier power spectrum method for fractal texture segmentation is given that is robust and usable in many situations. Errors at all stages are considered and presented as well as techniques to reduce, or make consistent these biases. These errors occur within the original due to the nature of the discrete Fourier transform, especially considering windowing edge conditions, curve fitting, and the addition of deterministic and non-deterministic noise elements.

Results are presented for synthetic fractal signals as well as on a real SAR captured image. Also presented are complete description and results for digitally calculating the Hölder order multifractal analysis that allow the creation of an extra set of measure images for segmentation.

1 Introduction

This paper considers some of the operational issues involved in fractal dimension image segmentation. The first half reviews the power spectrum method of extracting a fractal dimension from one and two-dimensional signals as well as a using a digital implementation of Hölder order multifractals.

These tools are tested using ideal synthetic fractal textures as well as real SAR images. Various windowing options, and full results demonstrate the mono- and multifractal nature in images that is extractable.

A final section gives a study on the stability of using these techniques as errors are introduced at many stages of the method. These included the problems of using a discrete Fourier transform and the log-log line fitting process. Also biases caused by discontinuities, noise and the addition of deterministic signals are considered. Because of these sources of error, the ability to easily use this technique outside of a research laboratory environment has to be treated with caution. Procedures for reliable analysis are given as well as areas of further research.

2 Fractal Segmentation

We first consider a mask method for calculating the fractal dimension and consider its use as a texture measure. Appendix A outlines the power spectrum method, for calculating the dimension of a signal, which requires matching an input signal to white noise filtered with the ideal Fourier space filter and extracting the spectral decay, β,

$$Q(k) = |k|^{-\beta/2}, \text{ where } \beta = 5 - 2D_F$$

for one-dimensional signals and,

$$Q(k_x, k_y) = |\mathbf{k}|^{\beta/2},\ |\mathbf{k}| = \sqrt{k_x^2 + k_y^2}, \text{ where } \beta = 8 - 2D_F$$

for two-dimensional images; where D_F is defined as the Fourier fractal dimension. We define the algorithm to calculate, via a fixed mask size, the fractal dimension for both one- and two-dimensional signals.

Given a signal of n elements, we move a window of length w one element at a time. We therefore obtain $n - w + 1$ values of D_F. Within this window the Fourier transform is taken and the power spectrum is computed, $P(k) = \Re(k)^2 + \Im(k)^2$, where $\Re(k)$ and $\Im(k)$ are the real and imaginary parts of the spectrum respectively.

Given P_i represents a measured digital one dimension power spectrum, we wish to fit it to the expected form of a fractal power spectrum, $\hat{P}_i = c|k_i|^{-\beta}$, where c and β are positive constants. To calculate c and β, we can use the least squares approximation that minimises the error,

$$\xi = \sum_i (\ln \hat{P}_i - \ln P_i)^2 = \sum_i (C - \beta \ln |k_i| - \ln P_i)^2, \qquad \text{which gives us}$$

$$\beta = \frac{(\sum_i \ln |k_i| \ln P_i)(\sum_i 1) - (\sum_i \ln P_i)(\sum_i \ln |k_i|)}{(\sum_i 1)(\sum_i (\ln |k_i|)^2) - (\sum_i |\ln k_i|)^2} \text{ and } C = \frac{(\sum_i \ln P_i) + \beta(\sum_i \ln |k_i|)}{(\sum_i 1)}$$

as solutions for $\beta = 5 - 2D_F$ and $C = \log c$. We have deliberately left the sums without limits, and left $\sum_i 1$ in the expression, as we may choose exactly which part of the power spectrum to feed into the algorithm. So $\sum_i 1$ is equal to the number of samples in the fit.

As we use about half of the spectrum, when w is small we have few data points for the fit. Many authors, especially in the field of chaos theory, point out that large numbers of data points should be used for the computation of fractal and other dimensions.

Windowing Strategies It is possible to window each set of data when the Fourier transform is calculated to avoid the 'leakage' phenomenon. Different windowing options including rectangular, where no change is made to the data, Parzen, Hamming and Hanning can be applied. Given a signal f_i, we can apply a window of size w, defined by a function g_i. We then have the resulting values $f_i' = f_i g_i$ as a basis for computing the power spectrum. In Fourier space, we have,

$$F_k' = F_k \otimes G_k$$

where the capital letters denote the Fourier transform of the corresponding lower case letters and $\otimes$ is the convolution operator. The computed spectrum is a convolution of the spectrum of the unwindowed data with the Fourier transform of the window. Consider the current situation presented above, where the original data is multiplied by a rectangular window, and defined as,

$$g_i = \begin{cases} 1 & -w/2 \leq i \leq w/2 \\ 0 & \text{otherwise} \end{cases} \quad \text{and } G_k = \frac{w}{2\pi} \operatorname{sinc}\left(\frac{kw}{2}\right)$$

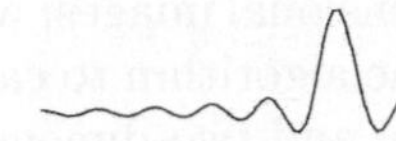

The power may 'leak' through the many lobes of the sinc function and alternative data windowing attempts to minimise this effect. It has been known that as there are generally discontinuities, of all orders, between the ends of the sampled signal, this gives rise to the computation of spurious high frequencies in the spectrum. This makes it clear that all windows should be even functions which fall off to zero at the extremities. This ensures continuity across the sample ends of the signal. The following windows, which have length $-w/2 < i < w/2$, all have obvious two dimensional equivalents,

Parzen or Triangular Window

$$g_i = \begin{cases} 1 - \frac{2|i|}{w} & |i| \leq w/2 \\ 0 & \text{otherwise} \end{cases}$$

Hanning Window

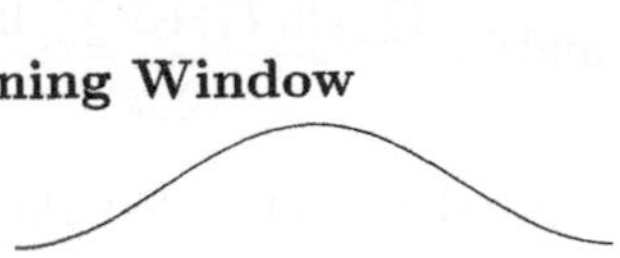

$$g_i = \begin{cases} \cos^2\left(\frac{\pi|i|}{w}\right) & |i| \leq w/2 \\ 0 & \text{otherwise} \end{cases}$$

This is the special case of a family of windows named after **Gersch.**

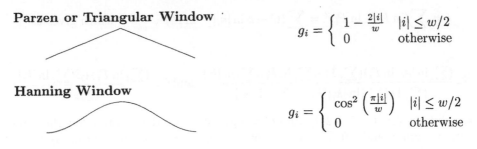

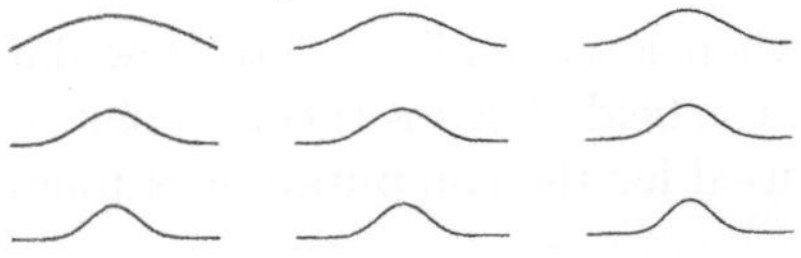

$$g_i = \begin{cases} \cos^n\left(\frac{\pi|i|}{w}\right) & |i| \leq w/2,\ n \in \mathbb{Z}^+ \\ 0 & \text{otherwise} \end{cases}$$

When $n = 1$ we get the cosine-tip window and for $n = 2$ we get the Hanning window. This window ensures continuity between the sampled signal ends up to the $(n-1)^{\text{th}}$ derivatives.

Hamming Window

$$g_i = \begin{cases} 0.54 + 0.46\cos\left(\frac{2\pi|i|}{w}\right) & |i| \leq w/2 \\ 0 & \text{otherwise} \end{cases}$$

The Hamming window, unlike the previous functions, does not quite reach zero at the sample signal ends.

Padding Strategies A final option to ensure that the resulting array is of the same size as the original data is a process of padding. For the one dimensional case an extra $w/2-1$ elements are added to the left and $w/2$ elements are added to the right of the original data. This can easily be extended for use with a two dimensional window with the logical extension by adding extra rows where necessary.

2.1 2-D Algorithm Using the 1-D Method

A method is given which uses the one-dimensional method described previously. As in the one dimensional approach a window of size w is chosen and is moved over the image row by row, moving one pixel at a time. For an input image of size $n \times n$ this generates an array of n rows by $n-w+1$ columns. For the next stage of the process the window is moved over the image column by column, which generates an array of $n-w+1$ rows by n columns. These two arrays are denoted, D^r_{ij} and D^c_{ij}, where the subscripts denote row and column respectively. An array of average D values, D^s_{ij}, can now be computed for a $w \times w$ square window by averaging the relevant w values from D^r and w values from D^c according to

$$D^s_{ij} = \frac{\sum_{k=i}^{i+w-1} D^r_{kj} + \sum_{l=j}^{j+w-1} D^c_{il}}{2w}$$

Hence a square array of size $n-w+1$ is generated which we assume will consist of numbers between 1 and 2 for a fractal image. It is noted that a synthetic Mandelbrot surface gives a fractal dimension for a grey level surface as a number between 2 and 3. Similarly to the one-dimensional case different windowing options can be applied: Rectangular, Parzen, Hamming or Hanning.

2.2 Direct 2-D Power Spectrum Algorithm

Given a two dimensional power spectrum, $P(k_x, k_y) = c|\mathbf{k}|^{-\beta}$, where $|\mathbf{k}| = \sqrt{(}k_x^2 + k_y^2)$, c is a constant and $\beta = 8 - 2D_F$. We can attempt to directly fit this function to the first quadrant of the measured two dimensional power spectrum. If we denote the measure spectrum by $P(k_x, k_y)$ and its best fit to a function of the form as $\hat{P}(k_x, k_y)$, then taking logarithms and calculating, the least squares approximation, gives us

$$\beta = \frac{N\sum_i\sum_j(\ln P_{ij})(\ln|k_{ij}|) - \left(\sum_i\sum_j \ln|k_{ij}|\right)\left(\sum_i\sum_j \ln P_{ij}\right)}{N\sum_i\sum_j(\ln|k_{ij}|)^2 - \left(\sum_i\sum_j \ln|k_{ij}|\right)^2}$$

and

$$C = \frac{\sum_i \sum_j \ln P_{ij} + \beta \sum_i \sum_j \ln |k_{ij}|}{N}$$

We have substituted N for $\sum_i \sum_j 1$, which gives the number of data points used in the least squares fit. The value depends on which of the following strategies is used. This method generalises in a straightforward way to higher dimensions. The only complication being the computation of triple sums for three dimensions, quadruple sums for four dimensions, and so on.

3 Test Signal Results

Synthetic aperture radar (SAR)[1, 5] images have been described as consisting of pure textural information and used for creating simulators [9]. First we check the performance of the algorithm on simulated pure fractal signals that use a mathematical description similar to the one used to detect the fractal dimension.

3.1 Simulated Synthetic Segmentation

We denote the true, or rather simulated, fractal dimension as D_F. The resulting fractal dimensions using the 1-D positive spectrum algorithm, excluding the DC level, is given in the following table. These used perfect synthetic periodic signals, created by Fourier smoothing white noise.

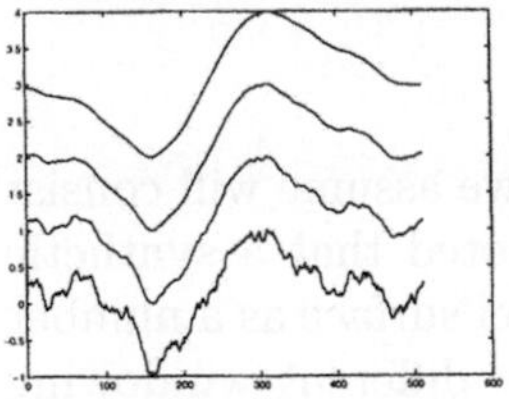

D_F	Rectangular	Parzen	Hanning	Hamming
1.3	1.250	1.388	1.213	1.210
1.5	1.458	1.425	1.424	1.421
1.7	1.667	1.637	1.635	1.632
1.9	1.876	1.848	1.845	1.842

The results are very accurate and varying the initial white noise makes very little difference to the magnitudes of the errors. As shown, the use of a windowing strategy to replace the rectangular window does not modify the accuracy substantially, although the results are reduced slightly. A windowing strategy is essential when the signal is not truly periodic, that is when there is a discontinuity at the ends. This effect is shown clearly in section 5.2.

Unfortunately, high accuracy comes with large window sizes. As the window size, w, is reduced, the accuracy of correct detection is substantially reduced. The following table lists the resulting calculated fractal dimensions when the real dimension $D_F = 1.3$. For very small masks the use of the non-rectangular windows becomes a liability as errors accumulate faster. For segmentation purposes, even if the correct absolute value becomes inaccurate, we are mainly concerned that there is a distinct difference between fractal dimensions.

w	Rectangular	Parzen	Hanning	Hamming
32	0.995	1.044	0.835	0.818
64	1.048	1.194	0.987	0.982
128	1.205	1.260	1.174	1.160
256	1.175	1.422	1.150	1.150
512	1.250	1.388	1.213	1.210

When the real dimension is increased to for example $D_F = 1.7$ the results are very similar.

w	Rectangular	Parzen	Hanning	Hamming
32	1.496	1.425	1.405	1.385
64	1.514	1.502	1.488	1.486
128	1.648	1.609	1.625	1.613
256	1.603	1.595	1.588	1.587
512	1.667	1.637	1.635	1.632

Overall, with experience, it is suggested that as large a mask as possible is used with iterative reduction when required. A rectangular windowing strategy has proved to be very durable and a Parzen window is favoured for its simplicity. As the window size reduces, the results appear to reduce and the signal appears smoother than it is. Again, it will be reiterated that in a real situation as the signals are not periodic within a window, one of the windowing strategies is essential and rectangular windowing should not be implemented.

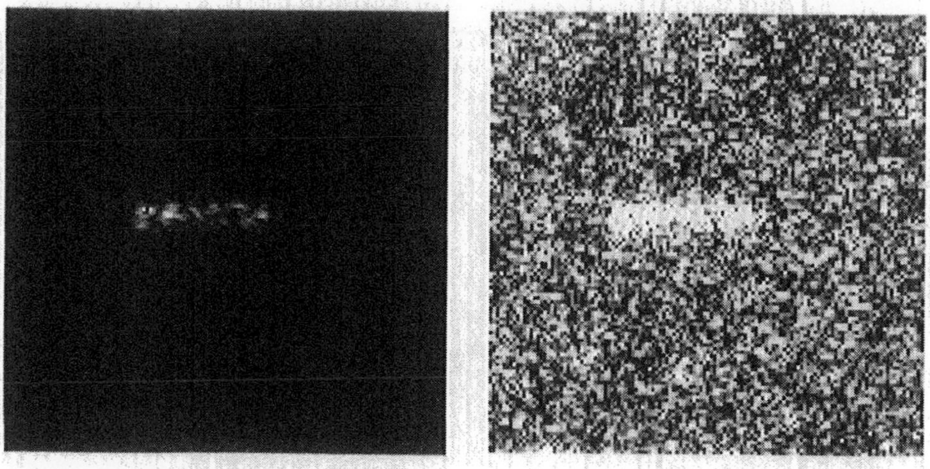

Figure 1. SAR image of the ship and magnitude gradient edge detector.

3.2 Real Fractal Segmentation

The image shown in figure 1 is a 128 × 128 section of a SAR image of a ship (central feature) taken in the English Channel and obtained with the SEASAT satellite, launched in June 1978 to carry out studies of the ocean surface using a 24 cm wavelength SAR. The magnitude gradient values show the speckle or radar 'clutter' caused probably by backscatter from the sea surface.

Careful use of the parameters in the Euclidean Canny edge detector (see Appendix B) shows that the grey levels directly can be used as a sensible measure resulting in usable segmentation, figure 2. It must be remembered that the ship is a 'Euclidean' object and if its characteristic grey level intensity is different enough from the more fractal-like ocean the previous grey level analyses will prove practical.

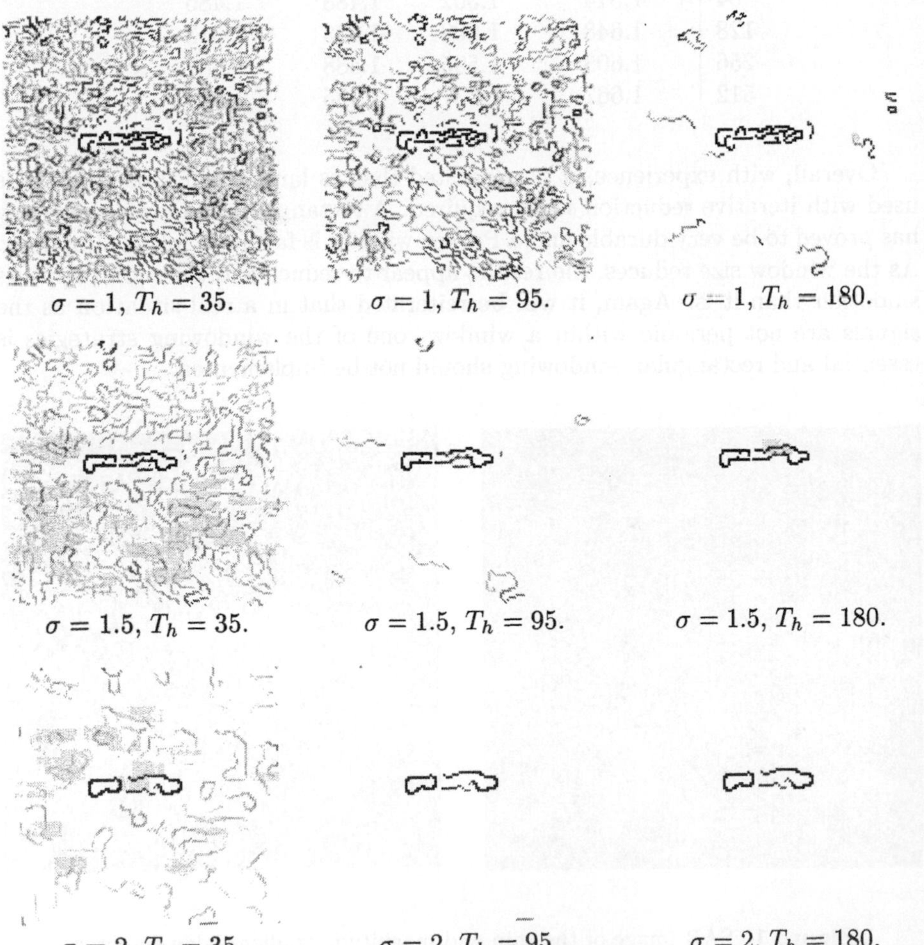

$\sigma = 1$, $T_h = 35$. $\sigma = 1$, $T_h = 95$. $\sigma = 1$, $T_h = 180$.

$\sigma = 1.5$, $T_h = 35$. $\sigma = 1.5$, $T_h = 95$. $\sigma = 1.5$, $T_h = 180$.

$\sigma = 2$, $T_h = 35$. $\sigma = 2$, $T_h = 95$. $\sigma = 2$, $T_h = 180$.

Figure 2. Canny filter varying standard deviation, σ and high threshold T_h.

We can now apply the Canny edge detector to the fractal measure image of the SAR ship image, figures 3–5. It is very easy to choose suitable parameters that extract a very similar enclosed shape. Both of these analyses have a distinct difference between the bow and the stern of the ship. One extra feature that is given by fractal dimension analysis is that an absolute measure for a region is given rather than just the rate of change. This can be demonstrated within the sea sections of the test image. The Canny edge detector created a mass of different edge lines which are filtered out completely at either the Gaussian function stage or the choice of threshold values. Using these techniques it is very difficult to understand or classify the sea areas. The fractal classification gives a numerical value that can be compared with other regions of seas in possibly other images. This means that if over a period of time a region's fractal nature changes, then this can also be observed. In the case of the sea this may relate to temperature or turbulence. So knowing the actual fractal dimension of a signal may have a direct relationship to some other property.

It is seen that the fractal dimension decreases substantially over the central region of the image which correlates with microwave backscatter from a man-made object. Clearly, a correlation between decreasing fractal dimension of SAR data and microwave backscatter from man-made objects could have useful applications. Further, for man-made objects with a high radar cross section, the energy of the return may in general be large, leading to high amplitude modulations in the SAR image. Since the parameter C is obtained in the least squares approximation, at the same time as we calculate β and D_F, we can easily extract and use the value of $C = \ln c$ as a classifier.

As a better classifier we can make use of the ratio D_F/C. This has the intuitive behaviour of having a high perceptive behaviour for man-made objects. This follows from the idea that man-made objects, as well as having a lower fractal dimension than natural ones, are likely to have a higher energy value. This means the ratio D_F/C is very small for man-made objects and very large for natural objects.

It is worth pointing out that the choice of windowing size could be very important to gain an accurate discrimination. Consider the case where there is an image that at a certain scale consists of two parts with different fractal dimensions. Using a window at this scale or finer will give the differentiation required. At fixed resolutions a discontinuity within a fractal dimension calculation when using the power spectrum method will cause an artifact. This was shown partially in the example with the SAR image of a ship.

An alternative proposed scheme is to believe that there are Euclidean style objects that are detectable and have a fractal nature within their structure. The procedure then involves carrying out a standard Euclidean segmentation process and then calculating the Fourier transform from a point set containing only the values within each segmented region.

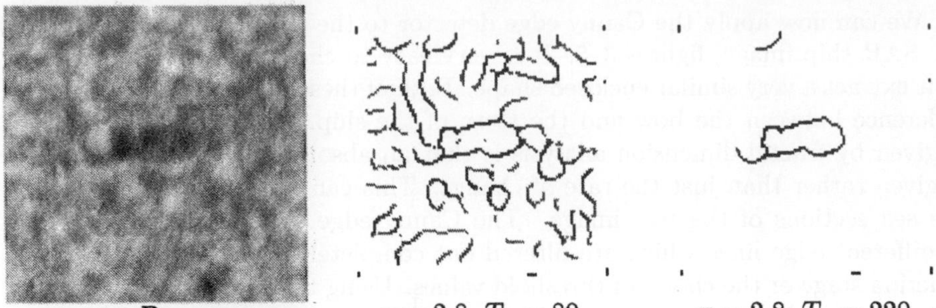

D_F $\sigma = 2.8,\ T_h = 80.$ $\sigma = 2.8,\ T_h = 220.$

Figure 3. Fractal segmentation followed by the Canny edge detection.

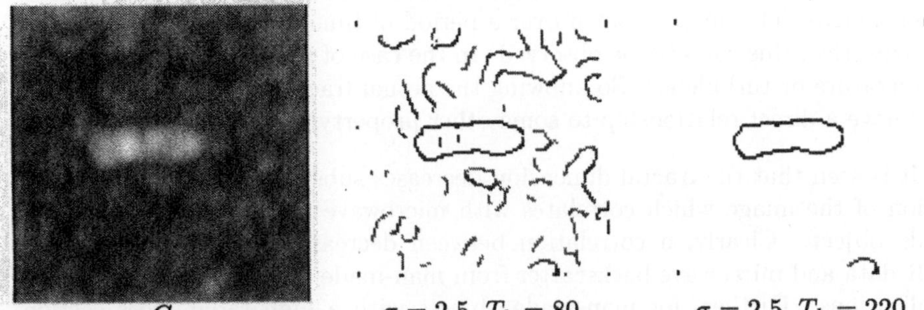

C $\sigma = 2.5,\ T_h = 80.$ $\sigma = 2.5,\ T_h = 220.$

Figure 4. Fractal energy segmentation.

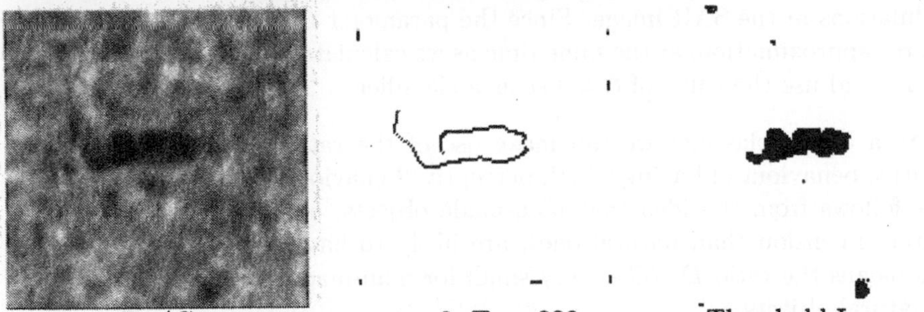

$D_F/C.$ $\sigma = 3,\ T_h = 220.$ Threshold Image.

Figure 5. Fractal dimension and energy segmentation.

4 Hölder Order Multifractals

We have now looked at a solution to the image texture segmentation problem. Jacques Lévy Véhel and Pascal Mignot at INRIA France have publicised a route that relies on the definition of Hölder continuity [14, 13, 12]. If we consider a function that has a Hölder continuous order, λ we have, a definition using the modulus of continuity,

$$w(\delta) = \sup_{|x-y|\leq\delta} |f(x) - f(y)| \text{ and } w(\delta) \leq k\delta^{\lambda} \text{ so } \lambda = \frac{-\ln k + \ln w(\delta)}{\ln \delta}$$

If we ignore the constant term and consider a two dimensional small neighbourhood around a point (x, y) then we can describe a digital coarse-grained local

Hölder order,

$$\lambda_{x,y}(i) = \frac{\ln c_{x,y}(i)}{\ln i} \qquad \text{and} \qquad \lambda_{x,y} = \lim_{i \to 0} \lambda_{x,y}(i)$$

where i is the size of the neighbourhood, and $c_{x,y}(i)$ is some capacity measure acting on the points within the neighbourhood. Now consider the two sets

$$E_\lambda = \left\{(x,y) \text{ s.t. } \lim_{i \to 0} \lambda_{x,y}(i) = \lambda\right\}, N_\epsilon^i(\lambda) = \text{card}\{(x,y) \text{ s.t. } \lambda_{x,y}(i) \in [\lambda - \epsilon, \lambda + \epsilon]\}$$

where card is the cardinality of the set. We now define $\varphi_h(\lambda)$ as the Hausdorff dimension of E_λ and then consider the double limit

$$\varphi_g(\lambda) = \lim_{\epsilon \to 0} \lim_{i \to 0} \frac{\ln N_\epsilon^i(\lambda)}{\ln i}$$

Each point in an image can then be described by a pair of values $(\lambda, \varphi(\lambda))$. φ_h measures the Hausdorff dimension of the set of points that have a certain λ, which gives a geometric description of the singularities. φ_g approximately defines the probability distribution of the singularities. We now consider a process that estimates φ_g with the following algorithm:

> **for** each i
>   **compute** all $\lambda_{x,y}(i)$
> **end for**
> $\lambda_{min} := \min\{\lambda_{x,y}(i)\}$
> $\lambda_{max} := \max\{\lambda_{x,y}(i)\}$
> **for** each i
>   **create** histogram with k cells of the $\lambda_{x,y}(i)$ values $[N_0^i, \ldots, N_{k-1}^i]$
> **end for**
> $\varphi_g(\lambda)$ calculated from a linear regression on $(\ln N_j^i, \ln i)$

The values λ_{min} and λ_{max} define the size of the i histograms so that they are comparable. The algorithm is only an approximation and for certain cases can yield unsatisfactory results. The number of cells in the histograms is arbitrary and does not take into account the ϵ limit in the definition of N_ϵ^i.

We have a choice of capacity measure $c_{x,y}(i)$; this can be related to the previously defined w_δ with the following, by considering W as the set of points within the neighbourhood around (x, y) that is defined by i:

$$\text{CDIF: } c_{x,y}(i) = \sup_{m,n,k,l \in W} |f_{m,n} - f_{k,l}|$$

or as an alternative centreing it in the middle of the neighbourhood,

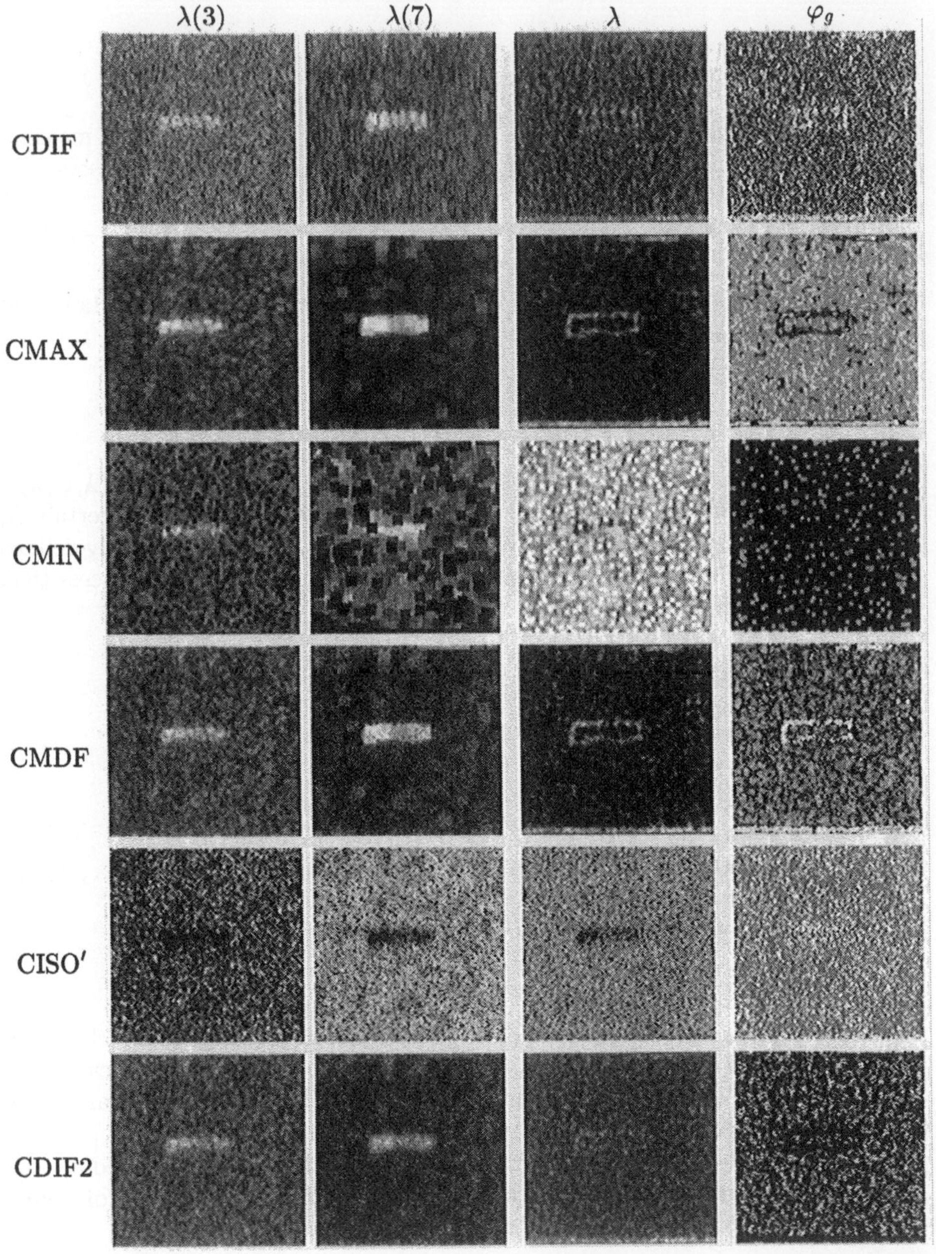

Figure 6. Hölder order multifractal analysis.

$$\text{CMDF: } c_{x,y}(i) = \sup_{m,n \in W} |f_{m,n} - f_{x,y}|$$

Véhel and Mignot [14] presented many other different capacity measures that can be used to detect and analyse different kinds of singularities, including;

$$\text{CMAX: } c_{x,y}(i) = \max_{m,n \in W} |f_{m,n}|, \qquad \text{CMIN: } c_{x,y}(i) = \min_{m,n \in W} |f_{m,n}|$$

$$\text{CISO: } c_{x,y}(i) = \text{card}\,\{(m,n) \text{ s.t. } f_{m,n} \equiv f_{x,y},\ m,n \in W\}$$

We now consider this process in practice. A series of square neighbourhoods, where i specifies the length of one side; then it is a simple process of creating the crude Hölder orders, $\lambda(i)$. The values of λ and φ_g are calculated via linear regression on the first few values of i. To show that alternative capacities can be developed, a maximum sum difference capacity is also shown,

$$\text{CDIF2: } c_{x,y}(i) = \sum_{m,n \in W} |f_{m,n} - f_{x,y}|$$

Different capacities detect different types of singularities; for example, the CISO capacity is ideal for detecting lines hidden within lots of noise. In this case a slight modification was applied that counted the number of 'similar' values,

$$\text{CISO}': c_{x,y}(i) = \text{card}\,\{(m,n) \text{ s.t. } |f_{m,n} - f_{x,y}| < 3,\ m,n \in W\}$$

Results for some of these capacity measures are shown in figure 6. All of these images can be used as measure images that require further segmentation and clustering algorithms. This as it stands is a first order digital approximation to the emerging field of Hölder order multifractal analysis. Within the last few years, Folconer and others have developed a strong mathematical basis to defining these quasi self-similar structures [4, 8].

5 Overview of Fractal Based Segmentation

Mandelbrot has stated that [7];

much of fractal geometry could pass as an implicit study of texture.

In the work reported here, a well tested algorithm has been applied for computing the fractal dimension of a fractal signal to synthetic aperture radar data amongst other examples. To include a note of caution, the stability and accuracy of any algorithm that extracts the fractal dimension from a linear regression can at times be very unstable for certain image types. A fractal segmentation technique must be considered as another tool to be used in conjunction with other image processing and understanding algorithms.

The application of random scaling fractal (RSF) models is not entirely appropriate to all forms of signal analysis and considerable effort has been put into classifying signals with genuine fractal characteristics. A more interesting general model has been proposed where the PSDF of a stochastic signal is assumed to be of the form (ignoring scaling)

$$P(k) = \frac{c|k|^{2g}}{(k_0^2 + k^2)^q}$$

This model encompass the fractal model and other stochastic processes such as the Ornstein-Uhlenbeck and Bermann processes. The parameters g and q could be used for texture segmentation on a wider class of signals and images than is appropriate for fractal dimension segmentation alone. Automatic extraction of parameters g and q is described in a previous publication [11].

It has been postulated that the fractal nature of a signal could be the sum of different fractal signals, vary over time or even be definable over a fractal timescale.

The current system that relates the spectral exponent β to a fractal dimension, as can be seen from Appendix A, is strictly applicable to only a subset of signal types, and defined in this paper only for fractional Brownian signals.

Although the principal of the algorithm is simple, and naïve implementation is not difficult, accuracy and acceptance within image processing has not been popular. Three anomalies are now presented to indicate some of the more subtle difficulties.

5.1 Errors with log-log plots

The power spectrum method is one of many different techniques that extract an approximation for the fractal dimension using a least squares fit from a bilogarithmic plot. We now present four methods that should all produce the same fractal dimension from a test signal.

Box counting One of the most popular algorithms for computing the fractal dimension of signals and images is the **box counting** method. Originally developed by Voss [15] but modified by others to develop a reasonably fast and accurate algorithm. Box counting in general involves covering a fractal with a grid of n-dimensional boxes or hyper-cubes with side length δ and counting the number of non-empty boxes $N(\delta)$. The slope β obtained in a bilogarithmic plot of the number of boxes used against their size then gives the fractal dimension (also known as the box or Minkowski dimension) where $D_B = -\beta$ (Figure 7). So the box counting measure gives us the generalisation

$$N(\delta) \propto \frac{1}{\delta^{D_B}} \quad \text{and} \quad D_B = \lim_{\delta \to 0} -\frac{\ln N(\delta)}{\ln \delta} \tag{5.1}$$

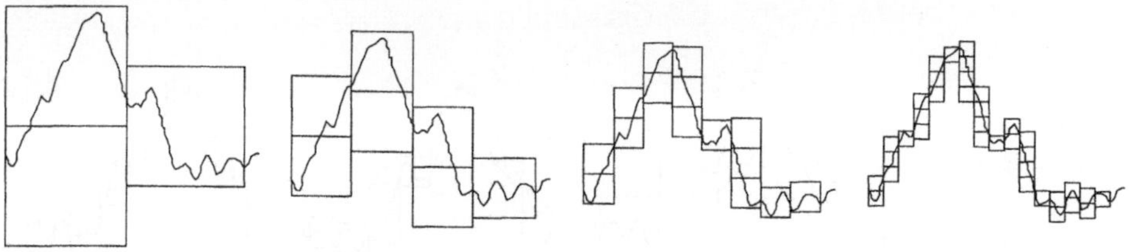

Figure 7. Box counting synthetic test signal; $D_B = 1.220$

Walking-Divider Introduced by Shelberg [10], this method uses a chord length (*Step*) and measures the number of chord lengths (*Length*) needed to cover a fractal curve. The technique is based on the principle of taking smaller and smaller rulers of size *Step* to cover the curve and counting the number of rulers *Length* required in each case. This approach is based on a direct interpretation of the above equation, where $N(\delta) \equiv Length$ and $\delta \equiv Step$ are estimated in a systematic fashion. It is a recursive process in which the *Step* is decreased (typically halved) and the new *Length* calculated. A least squares fit to the bilogarithmic plot of *Length* against *Step* gives the slope β where $D_L = -\beta$ (Figure 8).

$$D_L \approx -\frac{\ln N(\delta)}{\ln \delta}$$

The Walking-Divider method suffers from a number of problems. The initial and final *Step* must be carefully chosen. Shelberg described an appropriate starting value as half of the average distance between the points.

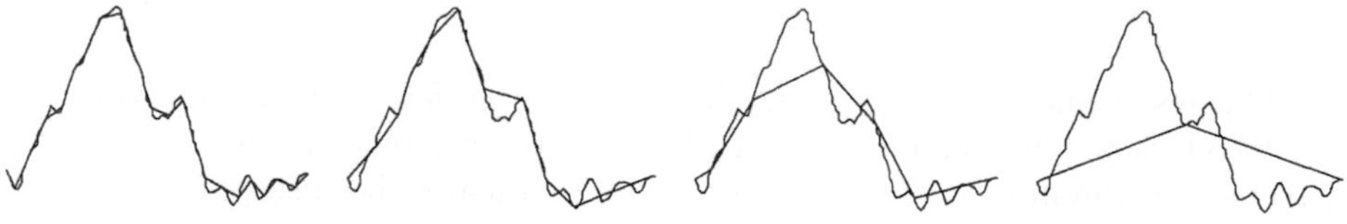

Figure 8. Walking-Divider strategy giving rise to the line dimension; $D_L = 1.309$.

Prism Counting Clarke [3] defines an algorithm based on the idea of box counting, in which instead of counting the number of boxes in a region for a given size, the area based on four triangles defined by the corner points is computed and summed over a grey level surface. The triangles define a prism based on the elevated corners and a central point computed in terms of the average of the four corners. A bilogarithmic plot of the sum of the prisms' areas for a given base area gives a fit to a line whose slope is β in which $D_P = 2 - \beta$ (Figure 9).

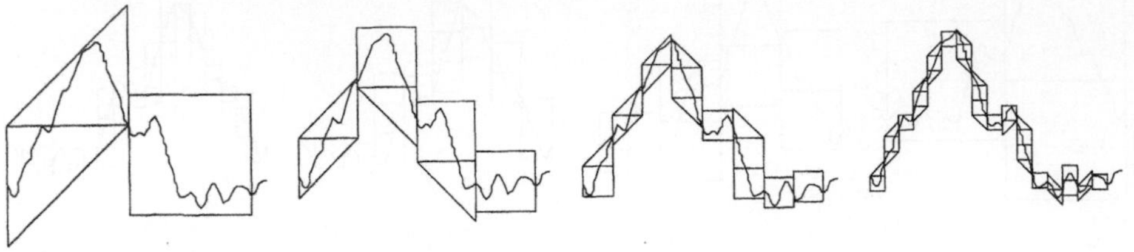

Figure 9. Two dimensional version of the prism counting strategy; $D_P = 1.239$

Epsilon-Blanket For curves we can consider the set of points whose distance from a curve is no more than ϵ. This gives a strip of width 2ϵ surrounding the curve. The length of the curve $L(\epsilon)$ can then be calculated directly or from the strip area $A(\epsilon)$ by $L(\epsilon) = A(\epsilon)/2\epsilon$. By using $L(\epsilon) \propto \epsilon^{1-D_\epsilon}$, D_ϵ can be computed (Figure 10).

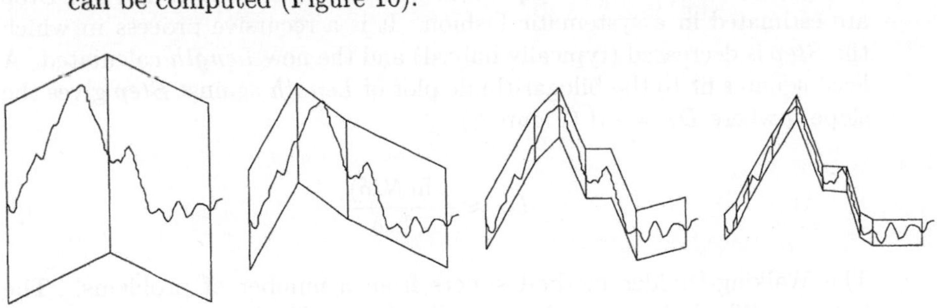

Figure 10. Covering strategy leading to the ϵ dimension; $D_\epsilon = 1.265$

The test signal in the previous figures has very few points, approximately 64, which is not uncommon in real life applications, but the resulting fractal dimension estimates vary from 1.2 to 1.3. This indicates the susceptibility during the least squares fit operation, and the fact that these methods are approximation to calculating the Haussdorf dimension. It is also a warning that the fractal dimension results for different methods, even similar ones, are likely to have biases. This means differences between two signals should be defined using the same method and that the method used should be stated in any results.

In general, box counting style algorithms behave well and produce accurate estimates for fractal dimensions between 1 and 1.5 for digital signals and between 2 and 2.5 for digital images and are easy to code and fast to compute.

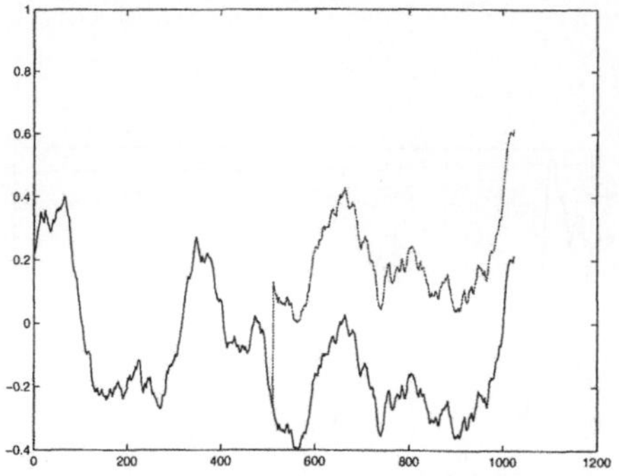

Original and Discontinuous Signals.

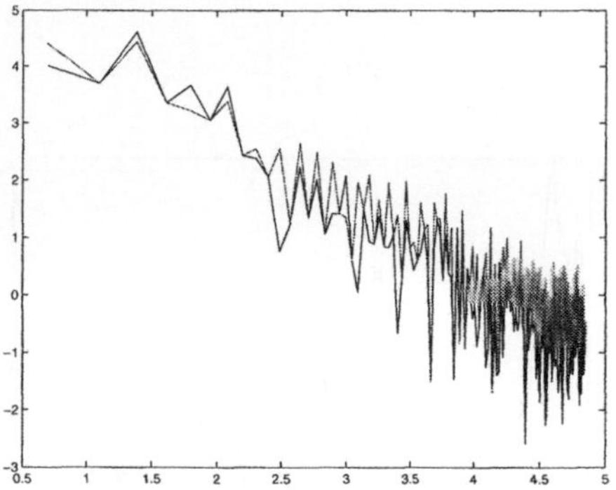

Log-log Power Spectrum.

Figure 11. Double discontinuity added lowering the spectral decay from; $\beta = 1.433$ to 1.233.

Outside this range (i.e. for higher fractal dimensions), they tend to give less accurate results; underestimating in most cases and saturating at an upper level [6]. The processes of fractional integration and differentiation have been shown to respectfully increase and decrease the fractal dimension of a signal in a controlled manner. It has been proposed to modify the signal so that the resulting detectable fractal dimension is first in an acceptable range.

5.2 Errors with Discontinuities and Noise

The original definition of the fractal dimension related to the spectral decay, β, defined in Appendix A, is specified for fractional Brownian motion. For other types of fractal signals this may not be appropriate.

To demonstrate this Figure 11 shows two signals, the second being equivalent to the first except a double discontinuity has been introduced. The original is synthetically created with a spectral decay, $\beta = 1.4$ and the value is extracted to within 1dp; $\beta = 1.433$. After two discontinuities are introduced a set of higher frequency component values are required that modify the fractal dimension extraction algorithm. The resulting spectral decay is lowered (i.e. higher supposed fractal dimension) to $\beta = 1.233$.

A similar problem with accuracy occurs when noise is added. Figure 12 shows four synthetic fractal signals that have been created with the same seed noise but have different fractal dimensions. Uniform noise is then added with a resulting signal-to-noise ratio (SNR) of 15:1. Spectral decays, β's are extracted, using the standard algorithm, before and after the noise is added. The following table lists the original and modified values for the β's.

β	1.2	1.3	1.4	1.5
No noise	1.1953	1.3047	1.4059	1.5146
Noise	1.2087	1.2913	1.3841	1.4780

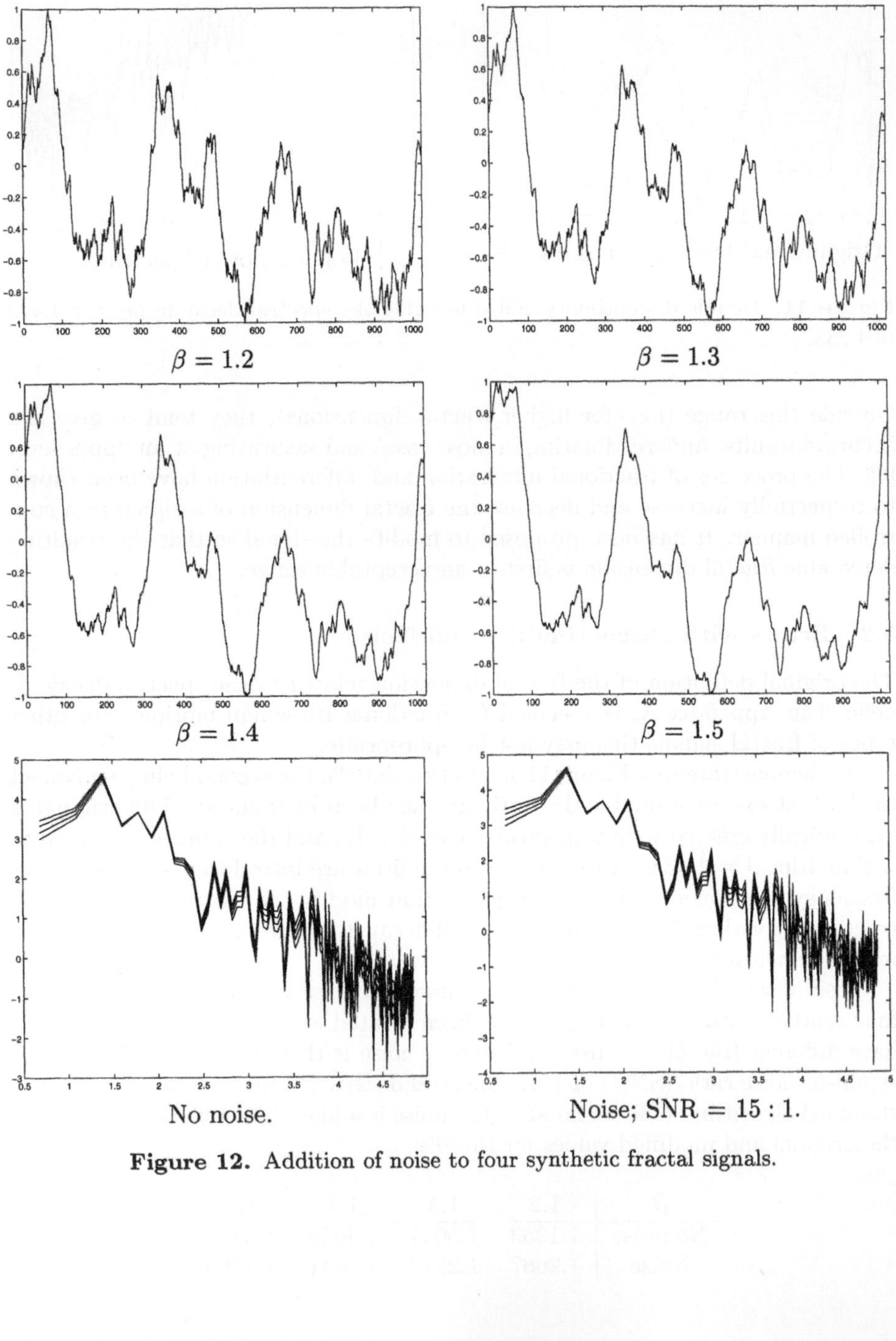

Figure 12. Addition of noise to four synthetic fractal signals.

As the table shows for the synthetic signals without noise we have an accurate extraction of the spectral decay. When there is a small amount of noise this affects the results in a noticeable way.

5.3 Errors due to Addition of other Signals

This basic random scaling fractal may be tailored by the addition of deterministic information. A standard method is to take a real function r for the deterministic information, Fourier transform to obtain a complex function, which we will denote $R + iH$, and then perform a normalisation of the two complex functions $R + iH$ and $N + iM$ so that the maxima of each are one. The complex deterministic and stochastic arrays may then be added together using a transmission coefficient t in the range $[0, 1]$ to give $G' + iH'$ where

$$G' + iH' = (1-t)(N + iM) + t(G + iH)$$

so that mostly stochastic information is used by setting t close to zero and mostly deterministic information is used by setting t close to one. The real and imaginary parts G' and H' are then filtered as before.

An alternative method can be applied by simply adding a deterministic signal with a fractal one. Given two signals $f_1(x)$ and $f_2(x)$ with fractal dimensions D_1 and D_2, the signal $f = f_1 + f_2$ will have a fractal dimension equal to $D = \max\{D_1, D_2\}$. To demonstrate that this is true for most cases, we know from the definition of self-similarity that

$$D_1 = \lim_{\delta\to 0} -\frac{\ln N_1(\delta)}{\ln \delta} \qquad \text{and} \qquad D_2 = \lim_{\delta\to 0} -\frac{\ln N_2(\delta)}{\ln \delta}$$

Now the new signal $f = f_1 + f_2$, with dimension D, will be covered by a maximum of $N_1 + N_2$ objects of size δ. Consider that $D_1 > D_2$ so $D_1 = \max\{D_1, D_2\}$ then

$$D = \lim_{\delta\to 0} -\frac{\ln\left(N_1(\delta) + N_2(\delta)\right)}{\ln \delta} = \lim_{\delta\to 0} -\frac{\ln N_1(\delta) + \ln(1+\alpha)}{\ln \delta} = D_1 - \lim_{\delta\to 0} \frac{\ln(1+\alpha)}{\ln \delta}$$

$$\text{where } \alpha = \frac{N_2(\delta)}{N_1(\delta)} = \frac{\delta^{D_1}}{\delta^{D_2}} = \delta^{(D_1-D_2)} \quad \text{and} \quad D = D_1 - \lim_{\delta\to 0} \frac{\ln(1+\delta^{\varepsilon})}{\ln \delta}$$

when $\varepsilon = D_1 - D_2$, and as $\varepsilon > 0$ and $\varepsilon < 1$ then as $\delta \to 0$ we know that $\ln(1+\delta^{\varepsilon})/\ln\delta \to 0$. Thus, as $\delta \to 0$, then $D \to D_1$ and therefore we have that $D = \max\{D_1, D_2\}$.

Figure 13 shows two samples of porous silicon captured from an SEM (Scanning Electron Microscope), and their log-log power spectrums. The complete image samples contain 256000 samples and 6000 frequency points, between two pre-specified frequencies, where used to create a least squares fit and extract the spectral decays, β. The two images represent similar samples exposed to different levels of UV light and the values of the spectral decay increase indicating that the roughness of the samples increases.

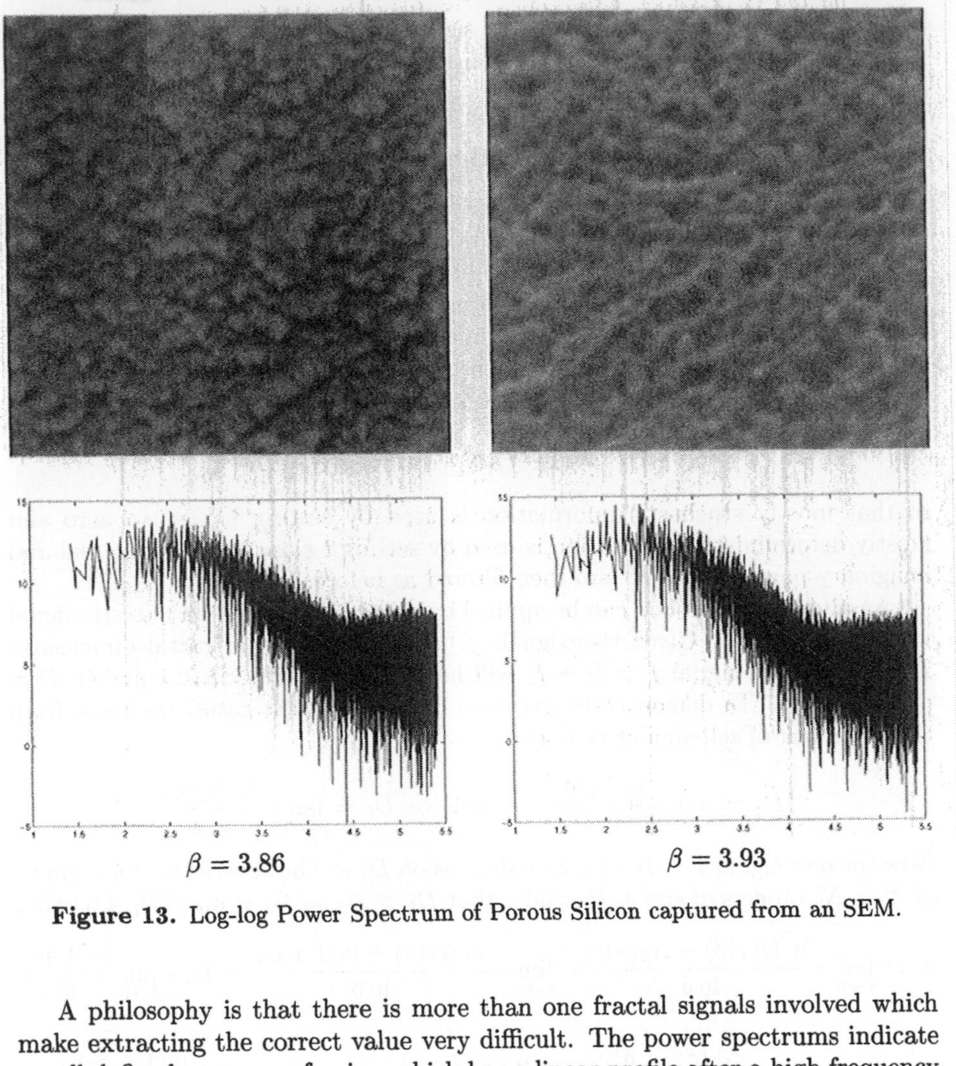

$\beta = 3.86$ $\qquad$ $\beta = 3.93$

Figure 13. Log-log Power Spectrum of Porous Silicon captured from an SEM.

A philosophy is that there is more than one fractal signals involved which make extracting the correct value very difficult. The power spectrums indicate a well defined presence of noise, which has a linear profile after a high frequency cut-off. A similar low frequency cut-off position can be detected that may correspond to larger structures and global intensity characteristics. Between these two cut-offs there is a definite spectral decay which is extractable and still has a few orders of magnitude.

If we believe the low and high frequency components are statistically consistent between samples the resulting spectral decay between these two frequency values will give an indicator of increasing roughness in the image and thus porosity of the silicon samples.

The SEM by its nature is a surface density measure and therefore fails to see certain features directly, most notably any concave structures that have been proposed to exist. This means for very porous materials the spectral decays may actually increase.

6 Conclusions

Fractal image segmentation techniques have been proposed many times for numerous different problems over the decades. Often these have not proved popular due to the extra processing required and the biases that can be introduced.

It has been shown that in the presence of additional noise, restricted sample sizes and floating point errors within the discrete Fourier transform and the log-log curve fitting resulting values can be wrong but have 'consistently' biased results. Therefore there is a recommendation for using a large number of data points, and a low signal-to-noise ratio to gain accurate results (higher than 2dp). When this is unavailable then comparative results can still be used, if all these factors and the fractal segmentation method used is fully specified. This would enable repeatability, and importantly comparability, in experiments as these factors cause biases rather than random errors to the results.

When looked at a macro statistics scale many image texture types can not totally be considered as monofractal in nature. Consideration of non monofractal statistics can become important and described in outline was an enhances fractal characteriser within the Ornstein-Uhlenbeck and Bermann process as well as a description of the Hölder order multifractal analysis.

All these techniques in the form used produce a single measure image that still requires further segmentation and clustering. A final comment is that these operators are usable, when treated carefully, and have groundings in physical mathematics, making them important and useful operator that should be in all image processing toolboxes.

Bibliography

1. B.C. Barber. Theory of digital imaging from orbital SAR. *International Journal of Remote Sensing*, 6(7), 1985.

2. J.F. Canny. Finding edges and lines in images. Technical Report 720, MIT, June 1983.

3. K. Clarke. Scale based simulation of topographic relief. *The American Cartogropher*, 12(2):85–98, 1988.

4. Kenneth J. Falconer. Dimensions and measures of quasi self-similar sets. *Proceedings of the American Mathematical Society*, 106:543–554, 1989.

5. J. Patrick Fitch. *Synthetic Aperture Radar.* Springer-Verlag, New York, 1988. ISBN: 038796665X.

6. J. Keller, S. Chen, and R. Crownover. Texture description and segmentation through fractal geometry. *Computer Vision, Graphics and Image Processing*, 45:150–166, 1989.

7. B.B. Mandelbrot. *The Fractal Geometry of Nature.* W.H. Freeman, Oxford, 1983.

8. Toby C. O'Neil. The multifractal spectrum of quasi self-similar measures. *Journal of Mathematical Analysis and Applications*, 211:233–257, 1997. Article No. AY975458.

9. T.K. Pike. Sarsim: A synthetic aperture radar system simulation model. *DFVLR-Mitt.*, 11, 1985.

10. M. Shelberg. The development of a curve and surface algorithm to measure fractal dimensions. Master's thesis, Ohio State University, 1982.

11. Martin J. Turner, Jonathan M. Blackledge, and Patrick R. Andrews. *Fractal Geometry in Digital Imaging.* Academic Press (Harcourt Brace and Company), 24-28 Oval Road, London NW1 7DX, June 1998. ISBN: 0-12-703970-8.

12. J. Levy Vehel, E. Lutton, and C. Tricot, editors. *Fractals In Engineering*, New York, June 1997. Springer-Verlag. ISBN 3-540-76182-9.

13. Jacques Lévy Véhel. Introduction to the multifractal analysis of images. In Yuval Fisher, editor, *Fractal Image Encoding and Analysis: A NATO ASI Series Book.* Springer Verlag, New York, 1996. held in Trondheim, Norway July 8-17, 1995.

14. Jacques Lévy Véhel and Pascal Mignot. Multifractal segmentation of images. *Fractals*, 2(3):379–382, 1994.

15. R. Voss. Random fractals: Characterisation and measurement. In R. Pynn and A. Skjeltorps, editors, *Scaling Phenomena in Disordered Systems.* Plenum, New York, 1985. ISBN: 0306421127.

A. Power Spectrum Dimension Extraction

A fractional Brownian motion (fBm), $B_H(t)$, is a function whose increments $\Delta B_H(x) = B_H(t+x) - B_H(t)$ have a zero-mean Gaussian distribution with variance given by

$$\langle |B_H(t+x) - B_H(t))|^2 \rangle \propto |x|^{2H}$$

The parameter $0 < H < 1$, defines its statistical scaling behaviour. We have that

$$\langle \Delta B_H(rx)^2 \rangle \propto r^H \langle \Delta B_H(x)^2 \rangle$$

The most famous case is when $H = 1/2$ and we have the familiar random walk associated with Brownian motion, and then $\Delta B^2 \propto \Delta t$.

If we consider the trace of an fBm covering a time period $\Delta t = 1$, and without loss of generality define the vertical range, $\Delta B_H = 1$. We know that $B_H(t)$ is statistically self-similar so if we divide the time span into $N = 1/(\Delta t)$ equal intervals, the vertical range within these intervals will be

$$\Delta B_H = \Delta t^H = \frac{1}{N^H} = N^{-H}$$

Using the box counting method, with boxes of length $\delta = N^{-1}$, the number of boxes required to cover each interval is

$$\Delta B_H \Delta t = \frac{N^{-H}}{N^{-1}} = N^{1-H} \Rightarrow N(\delta) = NN^{1-H} = N^{2-H}$$

and thus from the box counting equation 5.1,

$$D = 2 - H \tag{A.1}$$

So standard Brownian motion, which occurs when $H = 1/2$, has a fractal dimension of 1.5. This formulae can be expanded to cover signals of a higher topological dimensions, D_T, in a similar box counting manner with a resulting fractal dimension,

$$D = D_T + 1 - H \tag{A.2}$$

Now we come to a method favoured because it is generalisable and potentially more accurate computationally. This is the application of the Fourier power spectrum method. We consider the power spectrum of an ideal one-dimensional fractal signal with dimension D. This has the formula, $\hat{P}_i = c|k_i|^{-\beta}$, where c is a constant and β is the spectral exponent related to the Fourier transform dimension, D_F.

So fitting a least squares error line to the data (see section 2) the value of the spectral β, and hence D_F, can be found for the input signal. One of the main advantages of this approach is that the computation of D_F is based on an explicit formula. Note that the constant c provides a measure of the 'energy' of the signal since from Parceval's theorem (also called Rayleigh's theorem),

$$\text{Energy} = \int_{-\infty}^{\infty} | f(x) |^2 \, dx = \frac{1}{2\pi} \int_{-\infty}^{\infty} | F(k) |^2 \, dk = \frac{c}{2\pi} \int_{-\infty}^{\infty} \frac{1}{| k |^\beta} dk$$

We now wish to extract the relationship between the spectral exponent β and the Fourier dimension D_F. Consider a fractal signal $f(x)$ over an infinite support and a finite sample $f_X(x)$, given by

$$f_X(x) = \begin{cases} f(x) & 0 < x < X \\ 0 & \text{otherwise} \end{cases}$$

A finite sample is essential as otherwise the power spectrum diverges. Furthermore, in reality $f(x)$ is a random function and for any experiment or computer simulation we must necessarily take a finite sample. Let $F_X(k)$ be the Fourier transform of $f_X(x)$, $P_X(k)$ be the power spectrum and $P(k)$ be the power spectrum of $f(x)$.

$$f_X(x) = \frac{1}{2\pi}\int_{-\infty}^{\infty} F_X(k)\exp(ikx)dk, \; P_X(k) = \frac{1}{X}\left|F_X(k)\right|^2, \; P(k) = \lim_{X\to\infty} P_X(k)$$

The power spectrum gives an expression for the power of a signal for particular harmonics. $P(k)dk$ gives the power in the range k to $k+dk$. Consider next a function $g(x)$, obtained from $f(x)$ by scaling the x-coordinate by some $r>0$, the f-coordinate by $1/r^H$ and then taking a finite sample as before,

$$g_X(x) = \begin{cases} g(x) = 1/\left(r^H\right) f(rx) & 0<x<X \\ 0 & \text{otherwise} \end{cases}$$

Let $G_X(k)$ and $P'_X(k)$ be the Fourier transform and power spectrum of $g_X(x)$ respectively. We obtain an expression for G_X in terms of F_X,

$$G_X(k) = \int_0^X g_X(x)\exp(-ikx)dx = \frac{1}{r^{H+1}}\int_0^X f(s)\exp\left(-\frac{iks}{r}\right)ds$$

where we have substituted $s = rx$; hence

$$G_X(k) = \frac{1}{r^{H+1}} F_{rX}\left(\frac{k}{r}\right)$$

and the power spectrum of $g_X(x)$ is

$$P'_X(k) = \frac{1}{r^{2H+1}}\frac{1}{rX}\left|F_{rX}\left(\frac{k}{r}\right)\right|^2 \text{ and } P'(k) = \frac{1}{r^{2H+1}}P\left(\frac{k}{r}\right) \text{ as } X\to\infty$$

Since $g(x)$ is a properly scaled version of $f(x)$, their power spectra are equal, and so

$$P(k) = P'(k) = \frac{1}{r^{2H+1}}P\left(\frac{k}{r}\right)$$

If we now formally set $k=1$ and then replace $1/r$ by k we get

$$P(k) \propto \frac{1}{k^{2H+1}} = \frac{1}{k^{\beta}}$$

We have produced a signal that is statistically similar and is defined by the value H. The value of H has the exact same properties as that for an fBm. Now as $\beta = 2H+1$ we can calculate the Fourier fractal dimension D_F from Equation A.1, giving

$$D_F = 2 - H = 2 - \frac{\beta - 1}{2} = \frac{5 - \beta}{2}$$

The fractal dimension can be calculated directly from the above. This method also generalises in a straightforward fashion to higher dimensions. Thus we can define

$$\beta = 2H + D_\tau$$

and from Equation A.2 we have that $\beta = 5 - 2D_F$ for a fractal signal and $\beta = 8 - 2D_F$ for a fractal surface, or more generally,

$$\beta = 2(D_\tau + 1 - D_F) + D_\tau = 3D_\tau + 2 - 2D_F$$

and

$$D_F = D_\tau + 1 - H = D_\tau + 1 - \frac{\beta - D_\tau}{2} = \frac{3D_\tau + 2 - \beta}{2}$$

B. The Canny Edge Detector

In developing edge detecting filters two common problems are encountered. As the data elements are discrete it is not always obvious how to calculate a gradient value, and in the presence of noise many spurious edges can become apparent. In the mid 1980s the Canny edge detector [2] was presented to combat some of these problems and consists of three main stages.

1. Gaussian blur the image to reduce the amount of noise and remove speckles within the image. It is important to remove the very high frequency components that are higher than the following gradient filter as otherwise these will cause false edges to be detected.

2. Gradient detect using one of many standard filters, create two images, one containing the gradient magnitudes $\mathcal{G}$, and another containing the orientation $\theta(\mathcal{G})$. The most common implementations use a simple symmetric discrete first order derivative.

3. Threshold the gradient magnitudes, above a certain minimum threshold value, so that only major edges are detected. As well as this minimum low threshold value, a high threshold value is also specified. On any connected line, at least one of the edge points has to exceed this high value. This removes small or insignificant line segments.

By controlling the standard deviation, σ, of the Gaussian blurring operation, and the high, T_h, and low, T_l, threshold values, most general edges can be detected. If you know the type of edge you wish to detect and the type of noise that is present in the image then an alternative filter can be applied instead of the Gaussian filter.

Fractal Basins of Attraction in the Inversion of Geomagnetic Data

Gordon R.J. Cooper

Departments of Geophysics and Geology, University of the Witwatersrand
Johannesburg 2050, South Africa grcooper@iafrica.com

Abstract

Measurements of the strength of the magnetic field of the Earth (generally made from an aircraft or helicopter platform) are used to discern the nature of the near-surface geology, since many rocks contain magnetic minerals and are therefore weakly magnetic. Interpretation of the observed field values involves the geophysicist providing an initial estimate of the geology which is then improved or refined using least-squares inverse methods. This paper shows that, providing the model response to the parameters under inversion (such as the depth or position of a geological structure) is non-linear, then the set of models that converge to a given minimum in the misfit surface displays self-similarity and appears to be fractal in nature. The fractal dimension of the resulting basin of attraction can be reduced by increasing the damping of the inversion process.

1 Introduction

The subsurface structure of the Earth can be determined by the use of geophysical methods, and as a consequence geophysics is widely used in the search for minerals of economic importance. Readily available geophysical equipment can measure the strength of the Earth's gravitational field to about one part in 10^6, and its magnetic field to about one part in 10^5, in field conditions. Changes in the density or the magnetic properties of the rocks perturb the appropriate potential fields and allow the changes to be detected.

The measurements made in the field are interpreted to yield a model of the subsurface. This is done using either forward or inverse modelling procedures. Forward modelling consists of the geophysicist changing the parameters that describe a geological model (depths, widths, densities, and other properties of anomalous structures) and then calculating the resulting geophysical response. This is then compared to the observed field data, and when the two are the same the model is deemed a possible (simplified) representation of the subsurface. In inverse modelling however, the geophysicist selects various model parameters which the computer modifies mathematically in such a way as to make the model response equal to the observed data. This paper discusses fractal aspects of the inversion process.

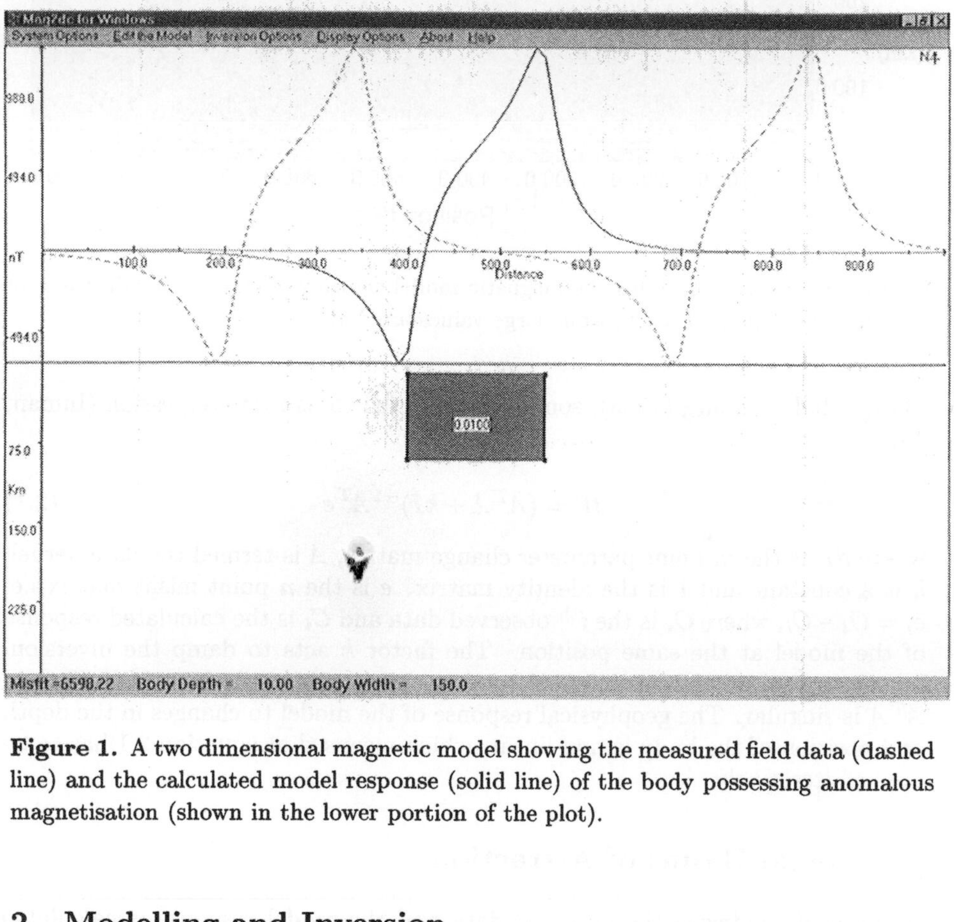

Figure 1. A two dimensional magnetic model showing the measured field data (dashed line) and the calculated model response (solid line) of the body possessing anomalous magnetisation (shown in the lower portion of the plot).

2 Modelling and Inversion

Figure 1 shows a simple geological model. The upper half of the plot shows the field data and the magnetic response of the model itself. The lower half of the plot shows the model, which in this case consists of a polygonal body with a susceptibility of 0.01 c.g.s. units. The rest of the subsurface is assumed to be non-magnetic at this stage. The model is two dimensional i.e. the body extends in and out of the figure to infinity in both directions, and the profile direction cuts the bodies being modelled at 90°. The magnetic anomaly of the body is calculated using the method of Talwani (Rasmussen and Pedersen, 1979).

The fit of the model response to the observed data will be best when the body is at either of two positions on the profile (200km or 700km) as there are two anomalies present in the observed data profile. The change in the model parameters from the initial starting set can be obtained from the technique of

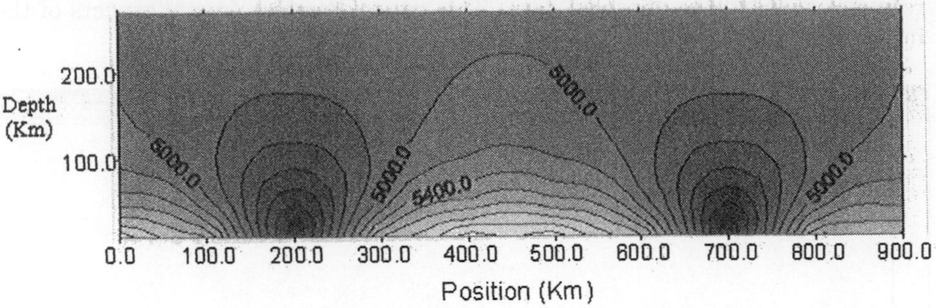

Figure 2. Misfit surface for the magnetic model shown in Figure 1. Small values of the misfit are coloured black, while large values are white.

Marquadt-Levenburg (1963), sometimes referred to as ridge-regression (Inman, 1975);

$$dP = (A^T A + kI)^{-1} A^T e \tag{2.1}$$

where dP is the m point parameter change matrix, A is termed the data kernel, k is a constant and I is the identity matrix. e is the n point misfit matrix i.e. $e_i = O_i - C_i$, where O_i is the ith observed data and C_i is the calculated response of the model at the same position. The factor k acts to damp the inversion, preventing the model parameters from assuming arbitrarily large values when $A^T A$ is singular. The geophysical response of the model to changes in the depth and position of the body is non-linear, which means that equation 2.1 has to be applied iteratively.

3 Fractal Basins of Attraction

If the misfit between the observed data and the model response is calculated over a grid of positions and depths, then the resulting image shows two minima, one at a position of 200km and the other at 700km. The inversion process described above requires an initial estimate of the model parameters, and this is then improved by the iterative application of equation 2.1. In this case the parameters under inversion were the depth and horizontal position of the body. The initial starting location for the body was then varied over a grid of positions and depths (10km apart initially), and the position that the body converged to was recorded. The result of the inverse process was that the body converged to one of the two minima in the misfit surface, or that it converged to neither. The latter case corresponded to the body position tending to $\pm\infty$.

The convergence of the body to either of the minima of the misfit surface as a function of the initial position and depth of the body was studied. Figure 3 shows the result of inverting the model from a grid of starting locations 10km

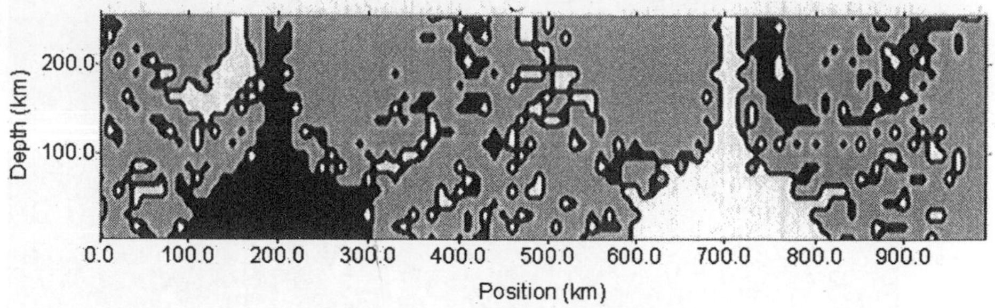

Figure 3. Inversion convergence results. The colour indicates the minimum that the inversion converged to ; black indicates the minimum at 200km, whereas white indicates that at 700km. If convergence to neither occurred, the plot is coloured grey. The grid of starting positions used were 10km apart.

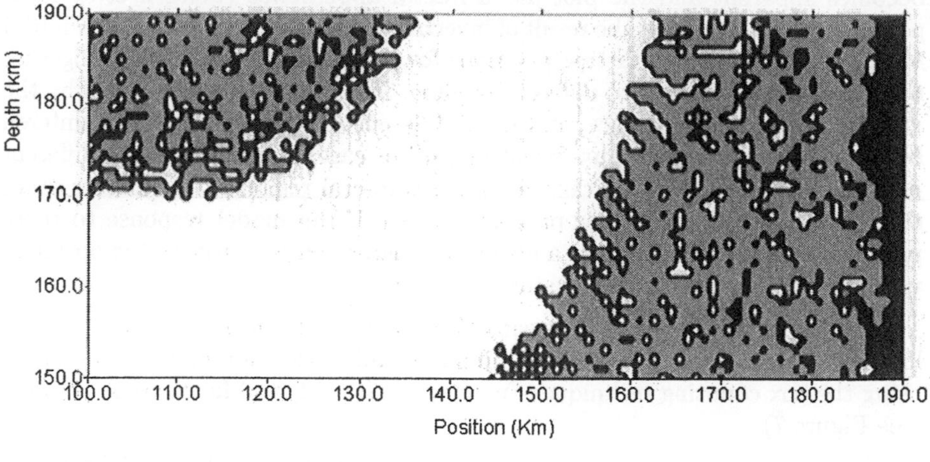

Figure 4. A portion of Figure 3 shown in more detail. The grid of starting positions used were 1km apart.

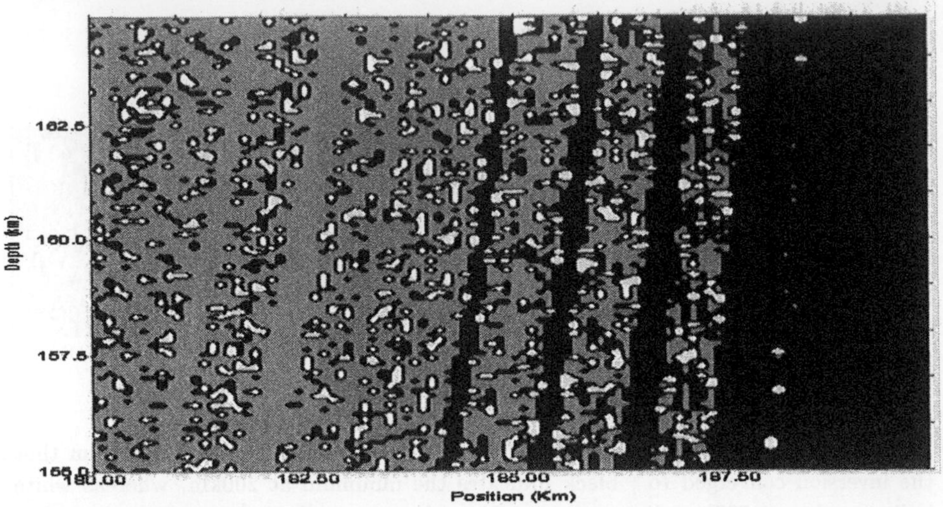

Figure 5. A portion of Figure 4 shown in more detail. The grid of starting positions used were 0.1km apart.

apart. In general the body converges to the nearest minimum, but the plot is not as simple as might have been expected. Figures 4 and 5 show portions of Figure 3 in more detail. As the plot is zoomed into, more and more detail becomes apparent and the plot has a fractal appearance (Peitgen et al 1992, p.63). Figure 6 compares the result of inverting the model from two nearby initial locations. Despite the starting positions for the inversion being close together the trajectories converge to different minima. The inversion is controlled by the gradient of the misfit surface, and small differences in the gradient encountered by the two trajectories add up, resulting (in this case) in convergence to different minima. Model parameters can only possess fractal regions in parameter space that converge to a particular parameter value if the model response to those parameters is non-linear, since a linear parameter will converge to the optimum value in one iteration from any starting value.

The fractal dimension of the basins of attraction (i.e. the set of all the starting locations that converge to a given minimum) for each minimum was measured using the box counting technique (Turcotte 1997, p.14) and found to be 1.8 ± 0.1 (see Figure 7).

The damping parameter k in equation 2.1 affected the form of the basins of attraction, and hence their fractal dimension. Figure 8 shows the result of increasing the value of k from 1.0 (as used previously) to 10.0. The fractal dimension decreased as the damping parameter was increased.

All calculations were performed using Turbo-Pascal for Windows real variables (6 bytes storage) and checked using Matlab double precision (8 bytes).

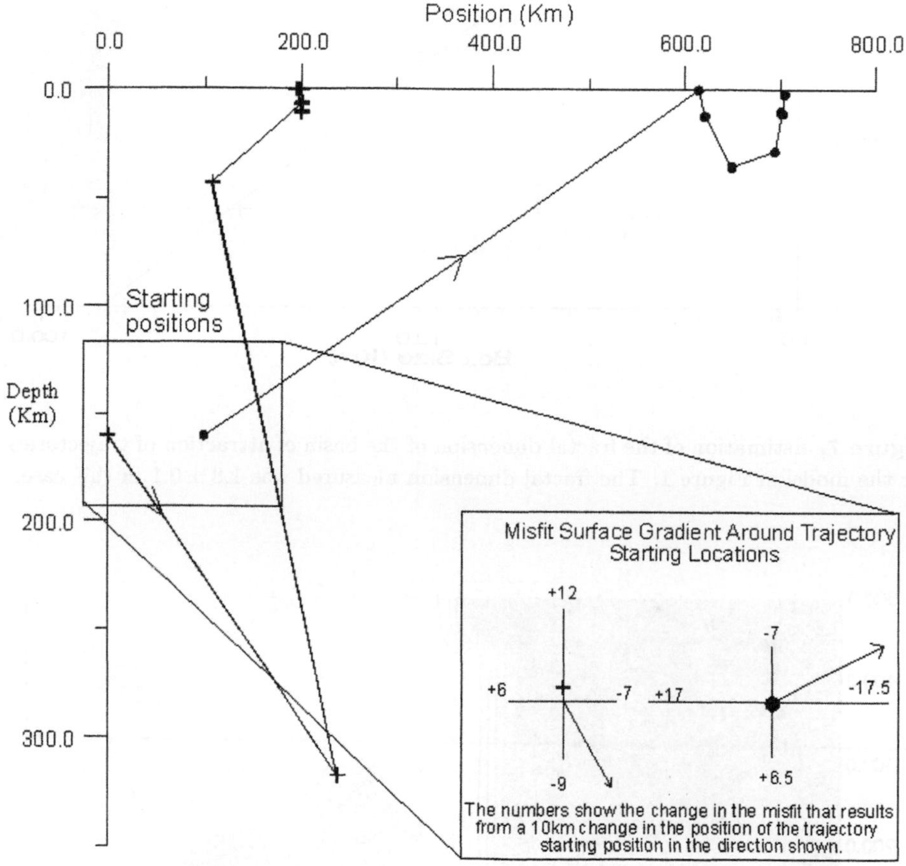

Figure 6. Behaviour of inversion trajectories that start from nearby locations. The gradient of the misfit surface at the starting points is shown in the box at lower right.

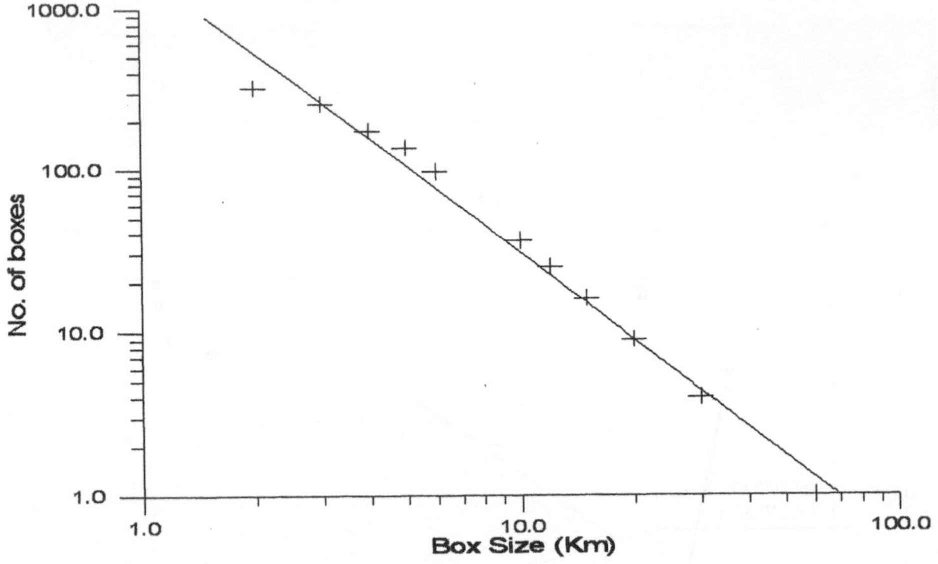

Figure 7. Estimation of the fractal dimension of the basin of attraction of trajectories for the model in Figure 1. The fractal dimension measured was 1.8 ± 0.1 in this case.

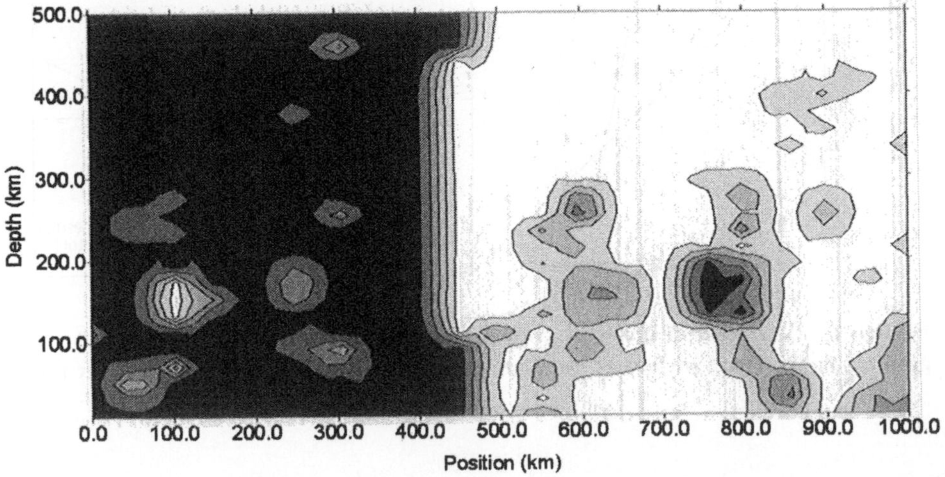

Figure 8. Inversion convergence results for the case when the damping coefficient in equation 2.1 is 10.0. The grid of starting positions used were 10km apart.

4 Three Dimensional Magnetic Modelling

A three dimensional magnetic model and dataset was now used in place of the two dimensional model described above. The dataset consisted of the anomalies from four dipolar bodies and was calculated from (Telford et al 1990, p.87);

$$F = \frac{m}{r^3}\left(3\cos^2\theta - 1\right) \tag{4.1}$$

where m is the magnetic moment of the dipole, r is the distance between the measurement point and the dipole centre, and θ is the angle between the dipole inclination and the vector from the measurement point to the dipole centre. The model consisted of the anomaly from a single dipole which had the same inclination as those used to generate the dataset. The starting locations for the inversion were located on a horizontal plane parallel to the ground surface, and initially at the same depth as the dipoles used to generate the dataset. The parameters under inversion were the position (EW and NS) and the depth of the dipole.

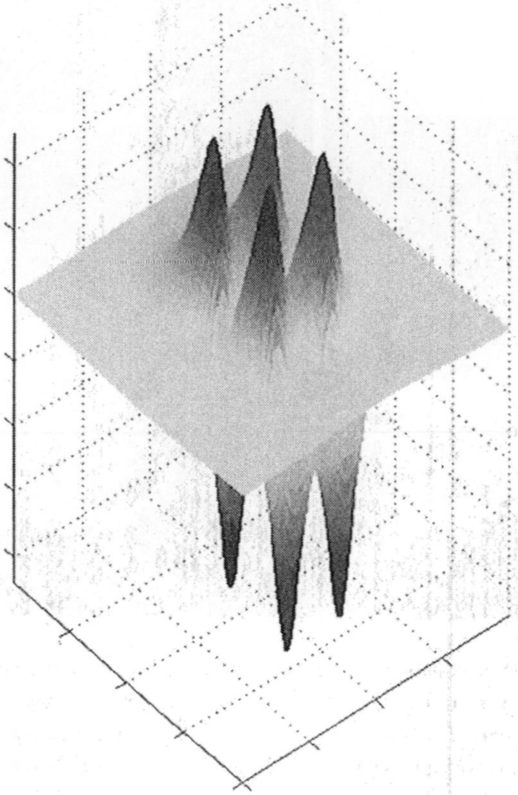

Figure 9. Induced magnetic anomaly generated by four dipoles with an inclination of 60° upwards measured from the horizontal. Magnetic intensity increases upwards.

Figure 9 shows the dataset used, and Figure 10 shows which of the four anomalies present in the dataset the model converged to under inversion. As in the two dimensional case, the pattern has a fractal appearance. Figure 11 shows a portion of Figure 10 in more detail. The fractal dimension of the basins of attraction was measured using the box counting method and a value of 1.45 ± 0.10 was obtained.

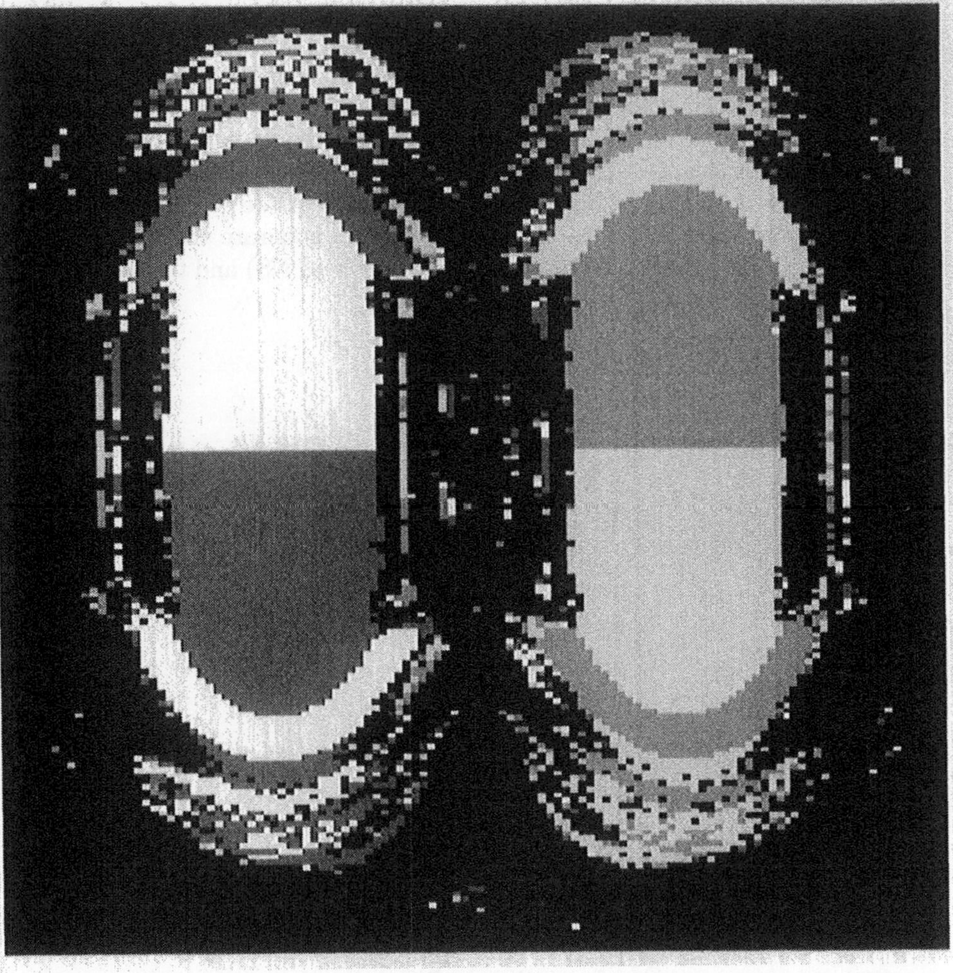

Figure 10. Inversion convergence results. The image shows a horizontal plane at a depth of 7.5 units, and it is coloured using different shades of grey according to which of the four dipolar features (shown in Figure 9) the inversion process converged to. If the model diverged to infinity then the image was coloured black.

All calculations were performed using Microsoft Fortran Powerstation 4.0 and used double precision (8 byte) accuracy.

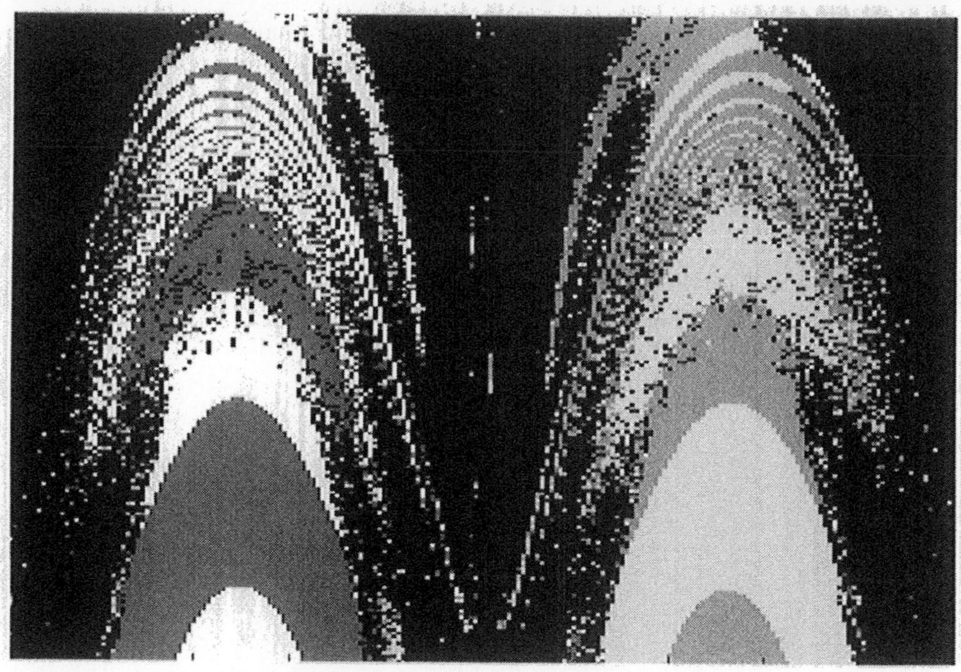

Figure 11. A portion of Figure 10 shown in more detail.

5 Conclusions

The set of model starting locations that converge under least-squares inversion to a given minimum in the misfit surface for two and three dimensional magnetic models displays self similarity and appears fractal in nature, providing that the forward model response to the parameters under inversion is non-linear. This makes the geophysical modelling of the subsurface geology more difficult than would have been expected, since it is harder to choose a suitable starting model for the inversion. The fractal dimension of the basins of attraction can be reduced by increasing the damping of the inverse process, though this has the disadvantage of slowing down the rate of convergence of the inversion.

Bibliography

1. Inman, J.R., 1975. Resistivity inversion with ridge regression. *Geophysics* v.50 , p.2112-2131.

2. Marquardt, D.W., 1963. An algorithm for least-squares estimation of non-linear parameters. *Journal of the Society of Industrial and Applied Mathematics* v.11, p.431-441.

3. Peitgen, H-O., Jurgens, H., and Saupe, D., 1992. *Chaos and Fractals, New Frontiers of Science.* Springer-Verlag New York, 984pp.

4. Rasmussen, R. and Pedersen, L.B., 1979. End corrections in potential field modeling. *Geophysical Prospecting,* v.27 p.749-760

5. Telford, W.M., Geldart, L.P., and Sheriff, R.E., 1990. *Applied Geophysics* 2nd ed. Cambridge University Press.

6. Turcotte, D.L., 1997. *Fractals and Chaos in Geology and Geophysics,* 2nd ed, Cambridge University Press, New York, 398pp.

Properties of Fractal Compression and their use within Texture Mapping

Martin J. Turner

ISS, SERC, Hawthorn Building, De Montfort University, Leicester LE1 9BH

Abstract

This paper gives a brief overview of fractal compression techniques using a quadtree decomposition based partitioned iterated function system, and applying it as a texture mapping process. To aid understanding of the process some visualisation techniques are presented, both 2-D and 3-D, as well as briefly considering compression performance and quality measures.

A description of the texture mapping process is given that maps 2-D images onto the surfaces of 3-D shapes, as well as the implementation of a fractal texture mapping prototype. Properties of fractal compression are presented that can be exploited within the texture mapping process. These include; *speed of creation, self-similarity* within common textures, *stability* of the algorithm given unknown initial conditions, *resolution independence* and *infinite resolution enhancement.*

Failings found in the prototype system involve the temporal nature of the image creation, and the addition of noise. Also, the transformations have a probabilistic distorting effect on the fractal decompression algorithm.

This technique is a valid operation if you accept certain assumptions; nature, or at least that part of nature that is used in texture maps, is fractal, and a 'good' fractal compression version of the image can be found. In these conditions the process is likely to be faster, take up less texture memory storage and be more pleasing to a human user.

1 Introduction

This paper gives a brief guide to fractal compression techniques for grey-scale images. Further information on all aspect of fractal compression, that has been referred to as self-referential Vector Quantisation, is available in the extensive literature [12, 19][1]. Section 3 considers certain coding artifacts that are apparent within fractal compression and the postulation that these are useful within a texture mapping operation. Then in section 4 a prototype system, previously first outlined and proposed in [23], is described that exploits these properties. The advantages and disadvantages of a fractal texture mapping approach are discussed and compared with mip-mapping techniques [13].

[1]The Leipzig Paper Collection on Fractal Image Compression is an extensive web based resource available at ftp://shear.informatik.uni-leipzig.de/pub/Fractal/papers/README.html

2 Fractal image representations

The main principle of fractal coding an image is the observation that self-similarity is found within images and is extractable. This self-similarity is definable in terms of affine transforms, w_i,

$$w_i \begin{bmatrix} x \\ y \end{bmatrix} = \begin{bmatrix} a_i & b_i \\ c_i & d_i \end{bmatrix} \begin{bmatrix} x \\ y \end{bmatrix} + \begin{bmatrix} e_i \\ f_i \end{bmatrix} = \begin{bmatrix} a_i x + b_i y + e_i \\ c_i x + d_i y + f_i \end{bmatrix}$$

The affine transform allows *rotation*, *shearing*, *translation* as well as *scaling*. An iterated function system (IFS) [1, 19, 2, 20] consists of a set of these transforms that makes up the whole image. The fundamental idea behind a fractal compression system is that if an image can be defined in terms of a self-similar set of transforms, and each transform can be described with an affine transform then a complete description of the image can be achieved by knowing only the numbers in the affine transforms. The infinite resolution intricate pattern shown below right is created with the following two functions

a_i	b_i	c_i	d_i	e_i	f_i
0.8	0.3	−0.2	0.9	-122	-7
0.1	0.5	−0.5	−0.4	51	526

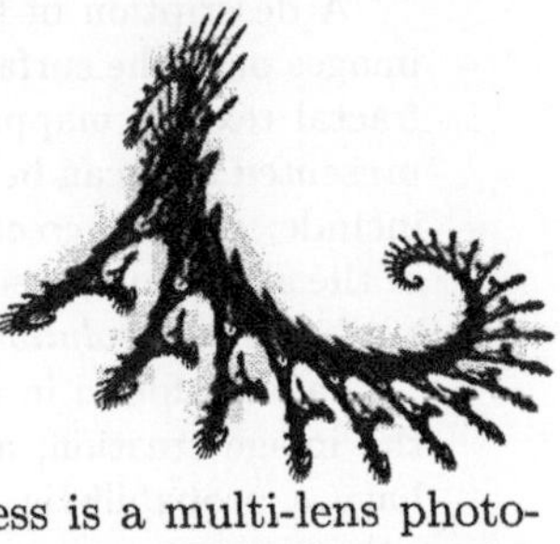

A common metaphor for this image creation process is a multi-lens photocopier with each affine transform describing one of the lenses. This photocopier is similar to a normal photocopier with the following extra properties:

- Each lens represents a copy of the original image which is transformed. This transformed image is allowed to overlap with any of the images of the other lenses.
- Each lens strictly reduces the size of the original image.
- The photocopier operates in an infinite loop, feeding back the output copy as the input to the next stage.

Creating the complete image IFS representation for an arbitrary image is a hard problem, and proven to be NP-Complete [22]. Two properties need to be defined, that of *contractive mapping* which specifies that a final image is always achievable and the *collage theorem* which shows how affine transforms can be combined. These two points will be briefly reviewed before a description of the partitioned IFS.

The **contractive mapping principle** ensures that there exists a unique fixed point known as an attractor. An affine transformation W is *contractive* on two points, x, y, if the distance, $d(\cdot)$, is of the form,

An 18 element IFS model spelling out an infinite number of *Martin* s,

a_i	b_i	c_i	d_i	e_i	f_i
0.03	−0.11	0.23	0.11	−6.3	4.4
0.06	0.11	−0.13	0.00	−6.9	5.6
0.06	−0.11	0.13	0.11	−5.0	5.0
0.03	0.11	−0.23	0.00	−5.4	5.0
0.03	−0.11	0.23	0.11	−2.8	4.4
0.03	0.11	−0.25	0.00	−3.1	5.0
0.08	0.00	0.00	0.11	−2.9	4.1
0.00	−0.11	0.22	0.00	−0.6	4.8
0.08	0.00	0.00	0.11	−0.8	6.0
0.00	0.11	−0.08	0.00	−0.6	6.2
0.09	0.00	−0.01	−0.13	−0.5	5.9
0.06	0.11	−0.13	0.00	−0.9	4.2
−0.12	0.00	0.00	0.16	1.8	5.6
0.00	0.11	−0.22	0.00	1.3	4.8
0.00	0.11	−0.25	0.00	3.2	5.0
0.00	−0.11	0.25	0.00	5.7	5.0
0.00	−0.11	0.25	0.00	7.3	5.0
0.08	0.11	−0.25	0.00	5.4	5.0

and a three element IFS model producing a double tailed spiral.

a_i	b_i	c_i	d_i	e_i	f_i
0.8	−0.4	0.2	0.9	106	84
−0.1	0.3	0.2	0.1	403	83
0.2	−0.2	0.1	0.2	365	94

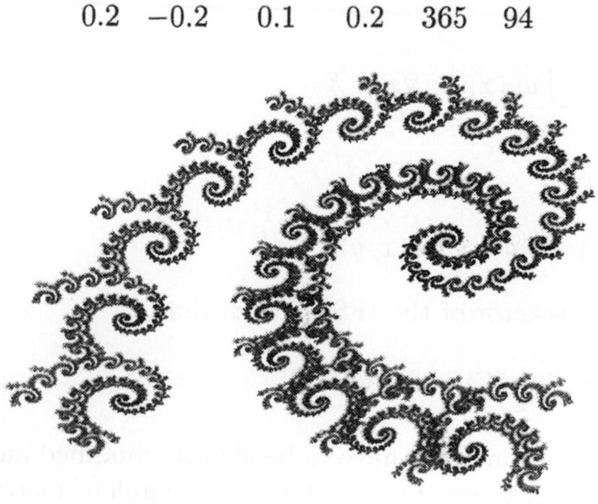

$$d(W(x), W(y)) < sd(x, y)$$

for some $s < 1$. The Contractive Mapping Fixed Point Theorem states *'If X is a complete metric space and $W : X \to X$ is contractive then W has a unique fixed point g'*. This can be proved using iteration. Given $x \in X$, create the sequence of points,

$$W^0(x) = x, \quad W^1(x) = W(x), \quad \ldots, W^i(x) = \overbrace{W(\ldots W}^{i}(x))\ldots)$$

Now given the distance relation,

$$d\left(W^{i+1}(x), W^{i+2}(x)\right) < sd\left(W^i(x), W^{i+1}(x)\right)$$

so that at each step the distance to the next point in the sequence is smaller than the distance from the previous point by a factor of $s < 1$. As geometric steps are taken and the space has no gaps, being a metric space, it must converge onto a single point, g.

$$g = \lim_{i \to \infty} W^i(x)$$

This fixed point is unique. Consider that there are two fixed points, x_1 and x_2, so $W(x_1) = x_1$ and $W(x_2) = x_2$. As W is contractive, then

$$d(x_1, x_2) < sd(W(x_1), W(x_2)) = sd(x_1, x_2)$$

but as $s < 1$ this inequality can not hold and therefore the fixed point $g = x_1 = x_2$ is unique for any initial value of x. For compression purposes one is looking for a contractive operator F whose fixed point $g = Fg$ is the best possible approximation of the original image. The *IFS Theorem* state that given a set of contractive transforms, $W = \{w_i : i = 1, \ldots, N\}$, with a transform defined as

$$W(x) = \bigcup_{i=1}^{N} w_i(x), \quad \forall x \in X$$

then when the contractive factor is s,

$$d(W(x), W(y)) \leq sd(x, y)$$

that has a unique fixed point, or attractor of the IFS, g, such that,

$$g = W(g) = \lim_{i \to \infty} W^i(x), \quad \forall x \in X$$

For point sets the normal distance functions have to be slightly modified and the Hausdorff function is a suitable representation. Given a complete metric space (X, d), the Hausdorff space, $\mathcal{H}$ where $\mathcal{H}(X)$ represents the space whose points are the compact subsets of X, excluding the empty set. The Hausdorff metric, h defines the distance between the sets A and $B \in \mathcal{H}(X)$, by

$$h(A,B) = \max\{d(A,B), d(B,A)\} \qquad \text{where}$$

$$d(A,B) = \max\{d(x,B) : x \in A\} \quad \text{and} \quad d(x,B) = \min\{d(x,y) : y \in B\}$$

where d is a standard distance function. Euclidean distance functions are commonly used,

$$d^p(x,y) = (|x^p - y^p|)^{\frac{1}{p}}$$

This is a measure that indicates in a general sense how similar two images are. To create large complex images say, T, from many transforms Barnsley [1] proposed the **collage theorem**. This states that given an IFS code $\{w_i, : i = 1, 2, \ldots, N\}$, with contractive factor $0 \leq s < 1$, then

$$h(T,g) \leq \frac{1}{1-s} h\left(T, \bigcup_{i=1}^{N} w_i(T)\right)$$

where g is the attractor of the IFS and $h(\cdot)$ is the Hausdorff metric. The collage theorem tells us that if in order for T and g to be close it is sufficient that T and $\bigcup_{i=1}^{N} w_i(T)$ are close. Now consider a *collage* or *encoding error*, $\epsilon_c = h(T, \bigcup_{i=1}^{N} w_i(T))$ and a *decoding error*, $\epsilon_d = h(T,g)$ then an upper bound for ϵ_d is

$$\epsilon_d = \frac{1}{1-s}\epsilon_c$$

For compression purposes the requirement is that the number of transforms, N, is to be as small as possible with ϵ_d smaller than a specified quality threshold level. The solution to this general problem has been shown to be very labour intensive to solve automatically. It was termed as the *Graduate Student Algorithm*, involving the following steps:

acquire a graduate student
place the student in a room with a picture and a computer
lock the door
    and wait ...
unlock when the picture has been reverse engineered

Arnaud Jacquin [16] in 1989 and later others [15] defined a flexible transform with the definition of a partitioned IFS (PIFS). This solved some of the limitations of the standard IFS. A PIFS relates one area of an image to another which is similar. Jacquin defined large areas called *domain blocks* that are associated with smaller areas called *range blocks*. Each range block is defined in terms of an affine transform and the location of the associated domain block.

The compression system involves splitting an image into, usually, a non-overlapping set of range blocks and then allocating an affine transform from a domain block to each range block. Each domain block needs to be strictly larger than the range block to make the system contractive. The coding process has the following two stages, an *encoding process* that involves searching for each range block a domain block that is closest in similarity and then encoding the affine transform and location, and a *decoding process* that starts with an initial image and applies the set of PIFS transforms repeatedly until the attractor is reached.

The general algorithm for all range-domain fractal coding methods is given below [2]. This creates the set $W = \bigcup w_i$. The only parameter to this algorithm is the *threshold value*, T that gives an indication of the quality required.

```
divide the image into a set of non-overlapping ranges R_i
mark all ranges as uncovered
while there exists an uncovered R_i
        choose domain D_i and map w_i s.t.
                distance := min(R_i − w_i(D_i))
        if (distance < T (Threshold) or size R_i < minimum)
            mark R_i as covered
            output transformation w_i and location D_i
        else
            partition R_i into smaller regions
            remove R_i from the list
        end if
end while
```

The key problem with the algorithm is the searching process, where the domain that gives the minimum distance is found. The order of complexity is multiplied by the number of possible domains, D_i, as well as the number of possible contractive affine transforms, w_i. Choosing a restrictive partitioning strategy and a simplified set of transforms reduces the complexity of the algorithm.

A slight modification is needed to the affine transform so that it can accommodate grey scale information. Two extra parameters, s_i and o_i, specify the *contrast* and *brightness* respectively.

$$w_i \begin{bmatrix} x \\ y \\ z \end{bmatrix} = \begin{bmatrix} a_i & b_i & 0 \\ c_i & d_i & 0 \\ 0 & 0 & s_i \end{bmatrix} \begin{bmatrix} x \\ y \\ z \end{bmatrix} + \begin{bmatrix} e_i \\ f_i \\ o_i \end{bmatrix} = \begin{bmatrix} a_i x + b_i y + e_i \\ c_i x + d_i y + f_i \\ s_i z + o_i \end{bmatrix}$$

To extract the contrast and brightness values from a domain ($D_i = \{\bigcup_{j=1}^{m} d_j\}$) and a range ($R_i = \{\bigcup_{j=1}^{n} r_j\}$), transform the domain block giving $D_i' = w_i D_i = \{\bigcup_{j=1}^{n} d_j'\}$, and then minimise the result,

$$\delta_i = D'_i - R_i = \sum_{j=1}^{n}(s_i d'_j + o_i - r_j)$$

by using the least squares approximation which gives

$$s_i = \frac{n^2 \sum_{j=1}^{n} d'_j r_j - \sum_{j=1}^{n} d'_j \sum_{j=1}^{n} r_j}{n^2 \sum_{j=1}^{n} d'_j - \left(\sum_{j=1}^{n} d'_j\right)^2}, \qquad o_i = \frac{\sum_{j=1}^{n} r_j - s_i \sum_{j=1}^{n} d'_j}{n^2} \quad \text{and}$$

$$\delta_i = \frac{\sum_{j=1}^{n} r_j^2 + s_i\left(s_i \sum_{j=1}^{n} (d'_j)^2 - 2\sum_{j=1}^{n} d'_j r_j + 2o_i \sum_{j=1}^{n} d'_j\right) + o_i\left(o_i n^2 - 2\sum_{j=1}^{n} r_j\right)}{n^2}$$

Jacquin introduced criteria for the s_i's, but in practice it has been shown that this can be broken with convergence still guaranteed [15, 18].

The choice of partitioning strategy is arbitrary, with any method that divides a range block R_i into a smaller non-overlapping set being acceptable. A common solution in the literature, and the one used here, is to use a quadtree division which divides each range block into four smaller quads. A technique which allows more arbitrary straight line divisions within the region has been proposed as well as triangular range blocks [12]. Optimal distortion based partitioning strategies have been studied by Hamzaoui and Saupe [14]. Each of these modifications allows more flexibility in the choice of domains and affine transforms, and thus they aim to increase the compression ratio without reducing the quality. In this specific partitioning case the range and domain blocks are all square, so the eight main transformations (rotations and reflections) were used.

3 Fractal Compression Properties

The contraction theorem states that the attractor will be reached irrespective of the starting point. This means that the initial image is in theory irrelevant. With small scaling factors, ($\leq 1/2$), the fixed point attractor is reached with only a few iterations. The first two generations for a PIFS description of the Mandrill image, are shown for different initial images (Grey level, Face, Checkerboard, Black and White bars, all Black and all White).

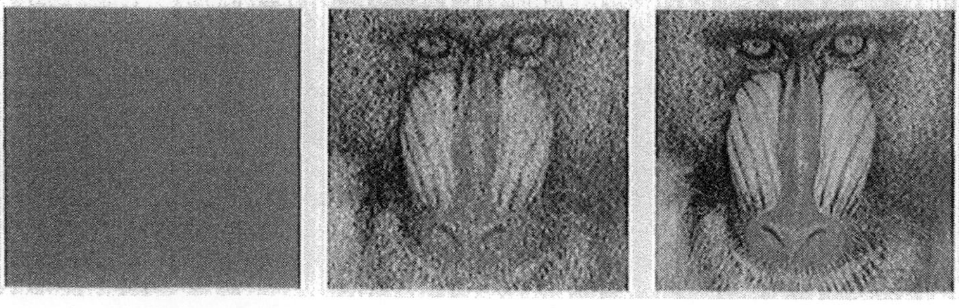

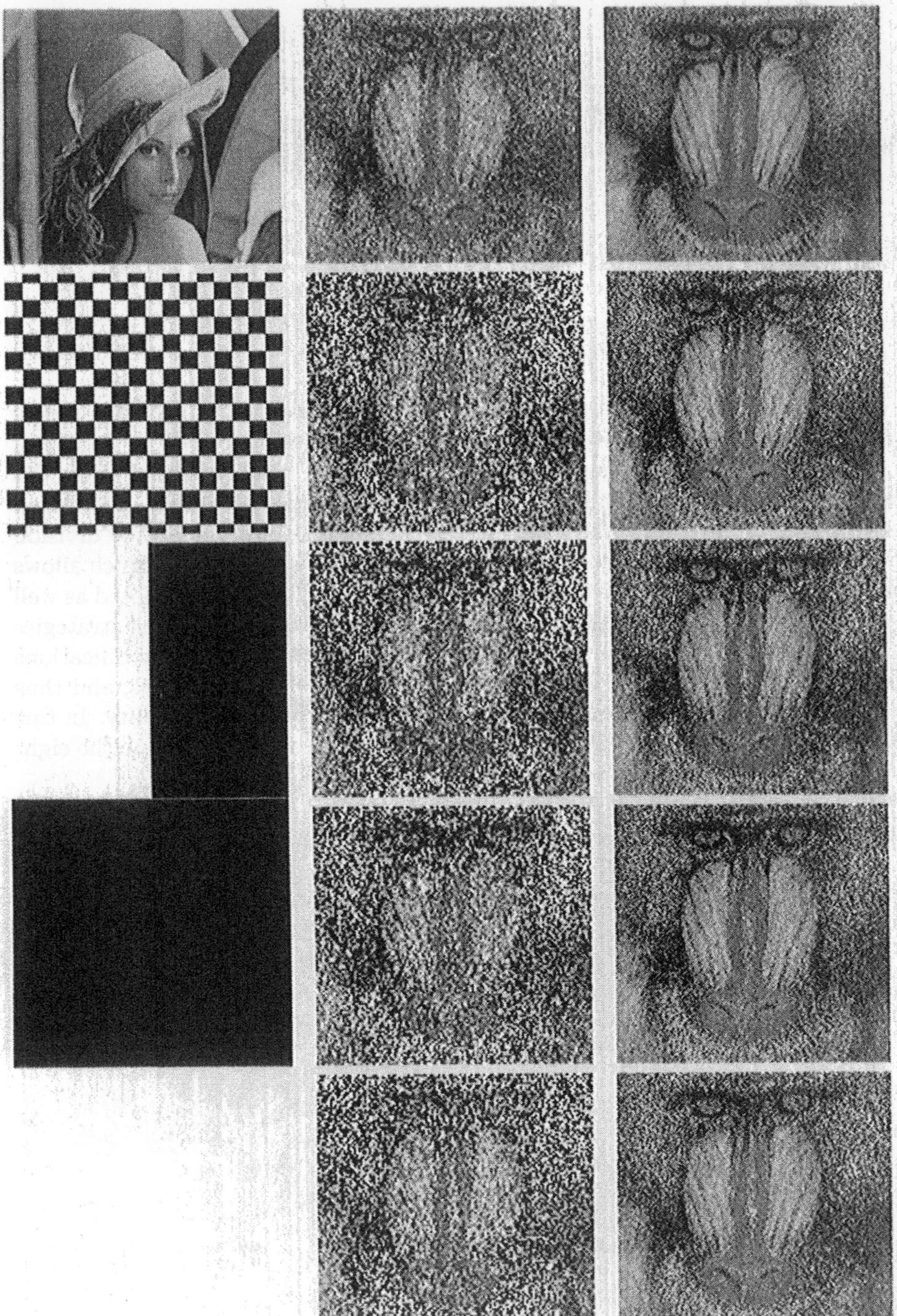

The final image can be seen to be converged to rapidly requiring only a few iterations. For the test initial images, the following graph shows the peak signal-to-noise ratio (PSNR) plotted for the first seven iterations.

$$\mathrm{SNR} = 10\log_{10}\frac{\langle f^2\rangle}{\frac{1}{MN}\sum_{j=0}^{N-1}\sum_{i=0}^{M-1}\left|f_{i,j}-\overline{f}_{i,j}\right|^2} = 10\log_{10}\frac{\sigma_f^2}{\sigma_{f-\overline{f}}^2}$$

$$\mathrm{PSNR} = 10\log_{10}\frac{\max\{f\}^2}{\frac{1}{MN}\sum_{j=0}^{N-1}\sum_{i=0}^{M-1}\left|f_{i,j}-\overline{f}_{i,j}\right|^2} = 10\log_{10}\frac{\max\{f\}^2}{\sigma_{f-\overline{f}}^2}$$

where f is the original image and $\overline{f}$ is the degraded image. As can be seen from the graph after six iterations the choice of the initial image was arbitrary.

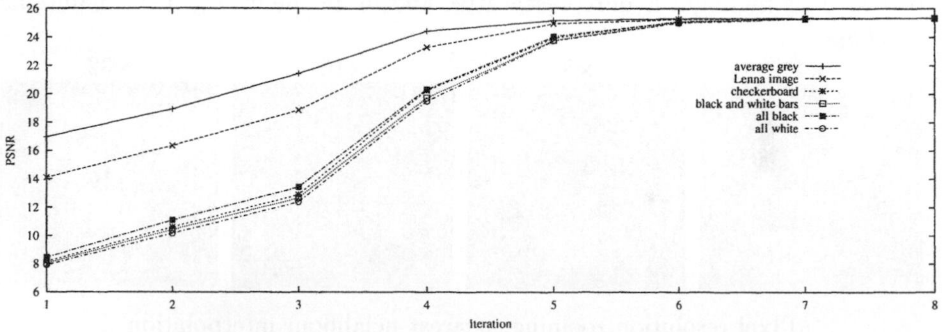

One of the main properties of an IFS is the concept of self-similarity. This means that an implicit definition of the image is given regarding itself and is not related to any particular resolution or specific number of pixels. The decoding process inherently, unless it is told otherwise, does not know, nor does it need to know, the resolution of the original image.

This gives a method of image interpolation where the decoder can create any size of image that is required. It is worth noting that when enlarging an image using this technique extra information is synthetically created from the definition of the whole of the image in some way. So, this is purely a type of interpolation and even if the created detail is appropriate it is purely generated. So, zooming in on an image of a wheat field will *not* show the structure of individual grains or stalks of wheat. This is an important point and has led to a slightly misleading claim with high performance image compression. Consider taking a grey scale image with resolution of 256×256, with one byte per pixel. This gives us an original image size of 65,536 bytes, before converting to a PIFS. If after converting the image the size of the PIFS is 6,554 bytes the resulting compression ratio is 10:1. Now if when decoding the PIFS it is magnified by a factor of 4 in both the horizontal and vertical direction, it will look like the original image resolution was 1024×1024. This gives us a total compression ratio of $10 \times 16 = 160$. Unfortunately, this extra information has been synthetically generated and, as mentioned above, even if it is appropriate there is no guarantee that it is correct.

This property of resolution independence gives us a process called *resolution enhancement.* The usual technique consists of converting the image to a very high quality PIFS and then decoding the image to a higher resolution before throwing away the PIFS code [19]. The process is shown for the Mandrill image with each image rendered at progressively higher resolutions. The final image has 1024 pixels (32×32) in the same area as the original had one pixel value. This process is compared with pixel enlargement using a nearest-neighbour interpolation scheme and an Overhauser cubic interpolator [7].[2] It is a subjective judgment which method produces the most pleasing, if not necessarily the most factually accurate result. The section of the test image shows the highlight in the pupil of the Mandrill's eye that due to limited sampling is represented as a few pixels. This highlight is unrepresentative and in all the enlargement maintains its shape.

×4 ×8 ×16 ×32

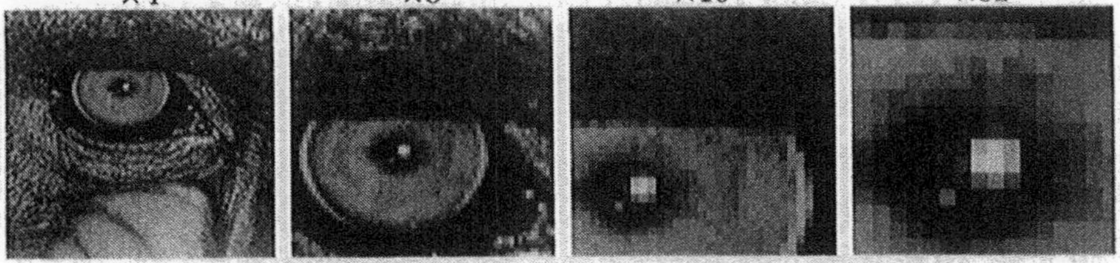

Pixel resolution zooming: nearest-neighbour interpolation.

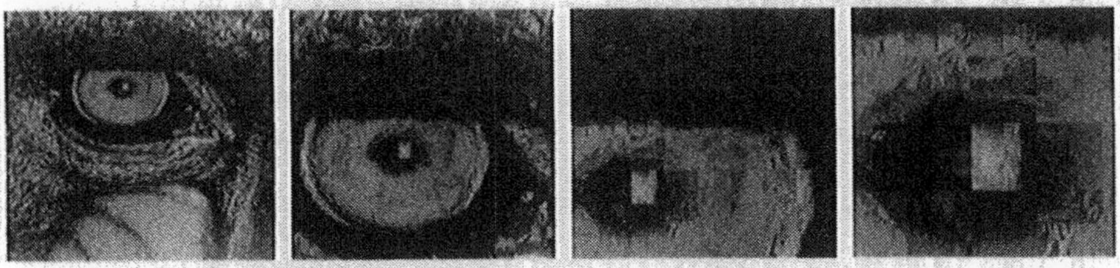

Fractal resolution enhancement.

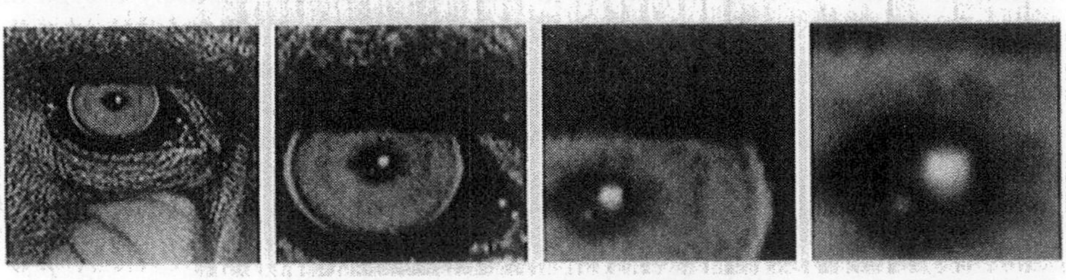

Overhauser cubic interpolation.

[2]The Overhauser cubic interpolant applied here was created using a slightly modified version of the IPE (Image Processing Environment) package written by Neil A. Dodgson.

For high resolution images created with fractal enhancement block edge effects, due to the partitioning strategy, are noticeable. To reduce this effect a post-processing filter can be applied along the boundaries. Given four points a, b, c, d in a scan line with a boundary edge between b and c, Fisher [12] recommend the following filter (other variations have been proposed that also include thresholds to avoid removing real edges [19]);

$$\begin{aligned} a' &= \frac{3a+2b+c}{6}, \\ b' &= \frac{2b+c}{3}, \\ c' &= \frac{b+2c}{3}, \\ d' &= \frac{b+2c+3d}{6} \end{aligned}$$

×50

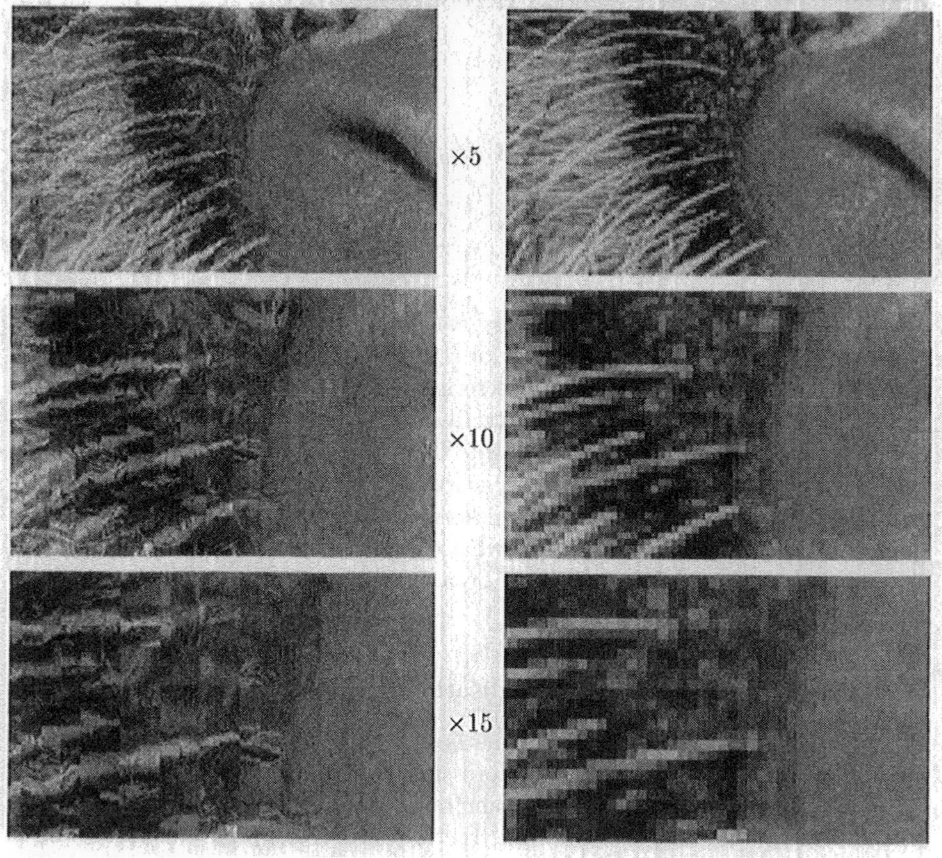

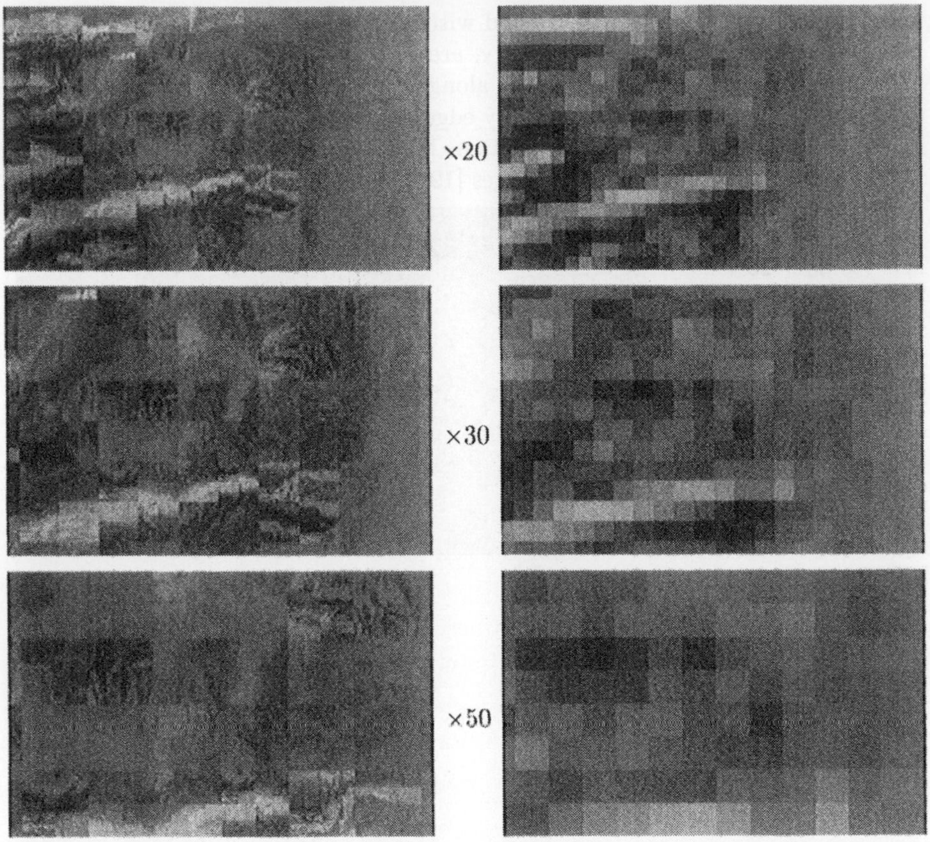

The previous series of images show resolution enhancement compared with nearest-neighbour pixel enlargement for a portion of the Mandrill's face around the whiskers on the right cheek. The series shows fractal enhancement up to a factor of 50 in both horizontal and vertical direction meaning that every pixel is reproduced as 2500 pixels (50 × 50). The images show two distinct texture types; the right hand side being very smooth while the left hand side has high contrast edges around the fur and whiskers. To demonstrate how information is 'interpolated' the whisker folical has what could be called fractal aliasing effects, that may not exist, causing there to be a spread of pixels rather than a sharp discontinuity.

3.1 Fractal Compression Visualisation

So far this paper has described a method that through partitioning tries to approximate a fractal IFS based compression system. The results, although not displaying increadably large compression ratios, do produce satistfactory images. In the following section the complete fractal texture mapping technique will be described and result presented. First the current PIFS process is visualised. The figure below shows the final fully quadtree partitioning grid for three different threshold values, T, on the Mandrill image ($T = 30, 25, 15$).

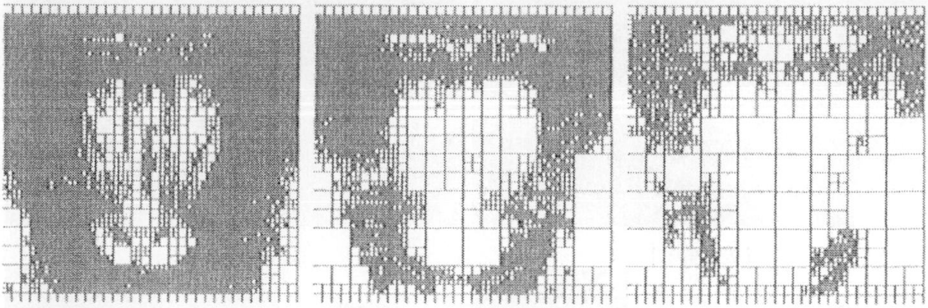

These images indirectly indicate, via the partitioning strategy, areas of high, middle and low complexity within the original Mandrill image. This can be used as a local complexity measure as it is very indirectly related to a local fractal dimension segmentation of the image. The boundary between a course and a fine partitioning area is one that may contain unwanted visual artifacts.

To visualise the process of image degredation and understand what is happening to the pixels, that constitute the image, a three-dimensional renderer can be applied. The following images, adapted from [24], shows fractal compressed images with different threshold values ($T = 1, 15, 22, 27, 30, 34$). In the top right of each image is part of the standard two dimensional view of the Mandrill image. The three dimensional view maps intensity values to height values so it shows how the coding process affects groups of pixels. The viewing point for the rendering is above the nose of the Mandrill looking just below the right eye.

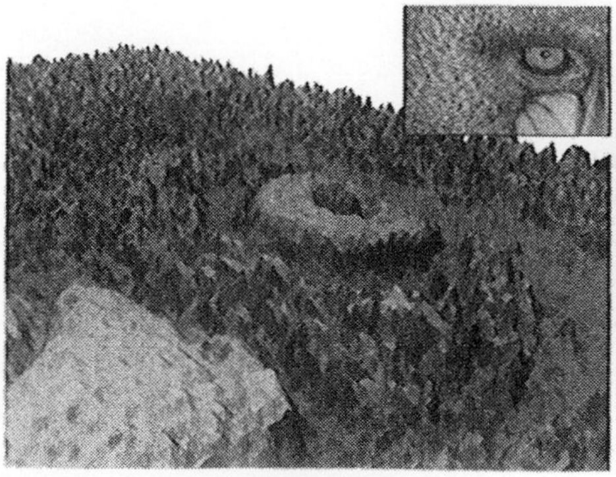

4.992 : 1

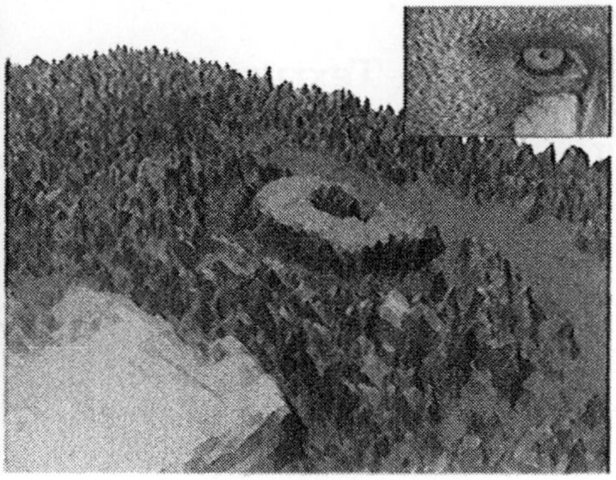

8.707 : 1

12.925 : 1

20.263 : 1

32.387 : 1

95.394 : 1

These images indicates how a computer vision system will see a decompressed image. It is imperative to consider this stage as more computer vision systems for object detection and recognition are by necessity using compression as a pre-step during the image acquisition and transmission stage. The overall shape of the structure is retained with blocking problems and removal of high frequency components occurring, at large threshold values. The last image has a very high compression ratio, nearly 100 : 1, but there are serious artifacts within the image visable both in the 2-D image and the 3-D computer visualisation.

4 Fractal Texture Mapping

The common process of texture mapping, in its basic form, applies a two dimensional image to the surface of a three dimensional shape. When the object is very close a higher resolution texture map is then required. Many rendering packages that create these images use a data structure technique called mip-mapping [26, 13]. This consists of a two dimensional texture image which has been repeatedly scaled by a factor of 2 to form a hierarchy, or pyramid, of texture maps. The texture map of the required resolution is then chosen. A pyramid data structure is a very efficient storage method for an image [5] and has two key features:

1. The total extra storage space requirement is minimal, as is the extra computational time required to apply any filters at all levels rather than just at the very fine level. If the original image is an $n \times n$ array where $n = 2^k$ then the number of pixels for the pyramid is

$$\begin{aligned} N_P &= \left(2^k\right)^2 + \left(2^{k-1}\right)^2 + \left(2^{k-2}\right)^2 + \left(2^{k-3}\right)^2 + \cdots \\ &= \left(2^k\right)^2 2^2 / \left(2^2 - 1\right) = \left(2^k\right)^2 4/3 = (4/3)n^2 \end{aligned}$$

2. For texture mapping there is an image available that is no more than 1.5× larger or smaller. These individual images at all the resolutions can also be optionally modified to obtain the best effect.

Now consider using a fractal description of a single two dimensional texture as a replacement for the standard mip-mapping texture. The technique is for the decoding process to transform domain and range blocks as required. The decoding or fractal texture mapping process can be considered as an infinite series of copy commands. Spot averaging over 4 points was used and considered adequate, although higher quality, with subsequently slower rendering times, can be achieved by averaging over more points. No special *initiator image* needs to be considered as the previous image at that location can be used. The texture mapping process is then applied using the following algorithm.

```
repeat forever
    for every range block
        map the range block onto image space
        map its domain block onto image space
        for every pixel in the transformed range block
            point average from the transformed domain block
        end for
    end for
end repeat
```

During the decoding process the range and domain blocks are calculated from the three dimensional transformations of the object that they are to be applied to. A fractal texture map is inherently resolution independent so only one representation would be required and that image can be stored in a compressed or semi-compressed format saving space. The image creation process is now proportional to the number of pixels required and the number of iterations. The progressive build up of the image means that for interactive use a crude image can be displayed when there is rapid movement, and a detailed view created for slow movement. This build up of quality over time makes efficient use of the graphics processing engine. Some of the main key advantages are:

- Fractal decoding is relatively fast, and implementable in hardware. In fact modern texture mip-mapping graphics processors carry out very similar operations.

- The quality of fractal coding degrades gracefully over quite a wide range of compression ratios.

- The fundamental operation within fractal decoding, that of copying one domain block to a range block, is very similar to any sampled texture mapping process currently being used.

- Systems with larger texture memory or more processing resources can use more detailed fractal map representations, and conversely systems with less texture memory can still use the same images by using a high compression ratio.

There are some potentially serious disadvantages in using fractal decoding techniques.

- Due to the perspective projection operation the property of convergence may be destroyed. It should be noted that a perspective projection as it has a singularity at the vanishing point can not be theoretically defined as an IFS. It is noted that the contractive mapping theorem can be broken for a set of affine transforms but a final image may still be converged to. The term 'eventual contractivity' has been used [18].

- Fractal encoding is potentially a very slow process. Fortunately, there are at present few occasions when new texture maps need to be recalculated, although in the future this may change.

- Any artifacts from the PIFS approximation that occur in the original fractal compression technique will still be present in the texture mapping process, and hence potentially noticeable in the persective projection. The blocking structure inherent in the quadtree partitioning strategy is the main cause of artifacts seen in this work. Very high resolution pixel based features are also likely to be detrimentally affected in the fractal encoding process.

- The sampling and averaging process is more complicated and if implemented naïvely may cause holes or edge effects to occur between range blocks. Over sampling may be required for special occasions, for examples with extremely distorted projections.

  The projection operation when applied to square texture maps has been shown to be non-optimal, which is a problem also seen when using the mip-mapping technique, that produces at times unavoidable blurring [25, 6]. Adjusting the shape or sample weights involved has been shown to alieviate these artifacts [3].

- Viewing only part of a texture map still involves creating a full image which could have excessive temporary memory requirements. If this is not done then severe artifacts including missing or distorted areas can occur.

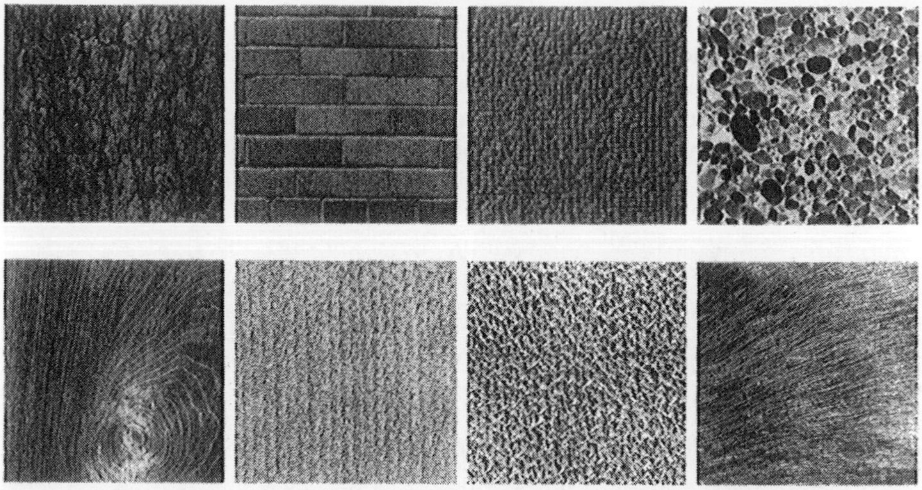

Fractals have been associated with the definition and construction of many synthetic procedural textures [8]. It is proposed that texture maps are then likely to be ideal candidates for fractal compression. In the following experiments eight Brodatz textures [4] were compressed (original scanned image resolution is 512×512 and consists of a section of the original photograph). To automatically analyse a recommended compression ratio that does not affect large scale features, the compression ratios are plotted against the SNR. This is shown below and for each line there is a change in gradient known as a *knee point.* It has been postulated by certain groups [21] that this is the ideal position to compress an image. After the knee point the image "quality" is slowly increasing as compression decreases and before the knee point the fractal coding affects large features in the image causing a rapid decrease in "quality". The term "quality" is quoted as it is an ill defined term and comparing it to SNR has been shown to be a very crude approximation [9, 10].

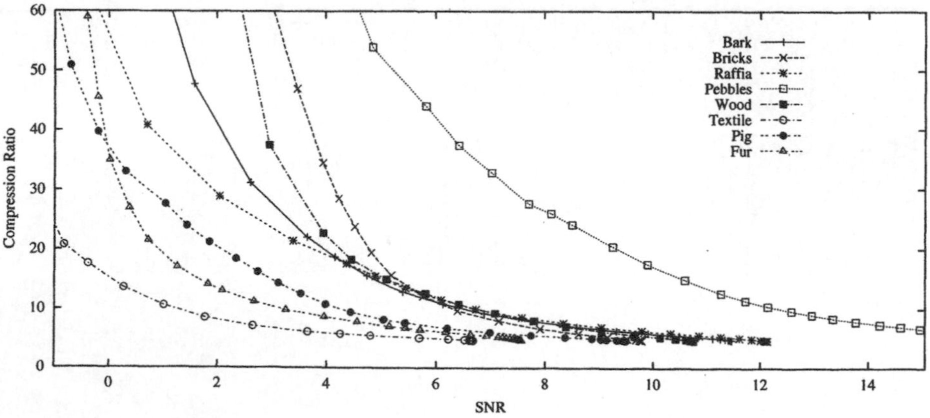

Choosing to take the fractal representation at about their knee points gives a reasonable compression ratio. For 15:1 this equates to one affine transform mapping on to an average of 22 pixels. The next stage is to confirm that aesthetically pleasing images are created when they are fractally enlarged. The eight image pairs below show a very small section of each texture image (128×40 pixels) where each pixel in the original image is now represented as 64 pixels in the enlarged version.

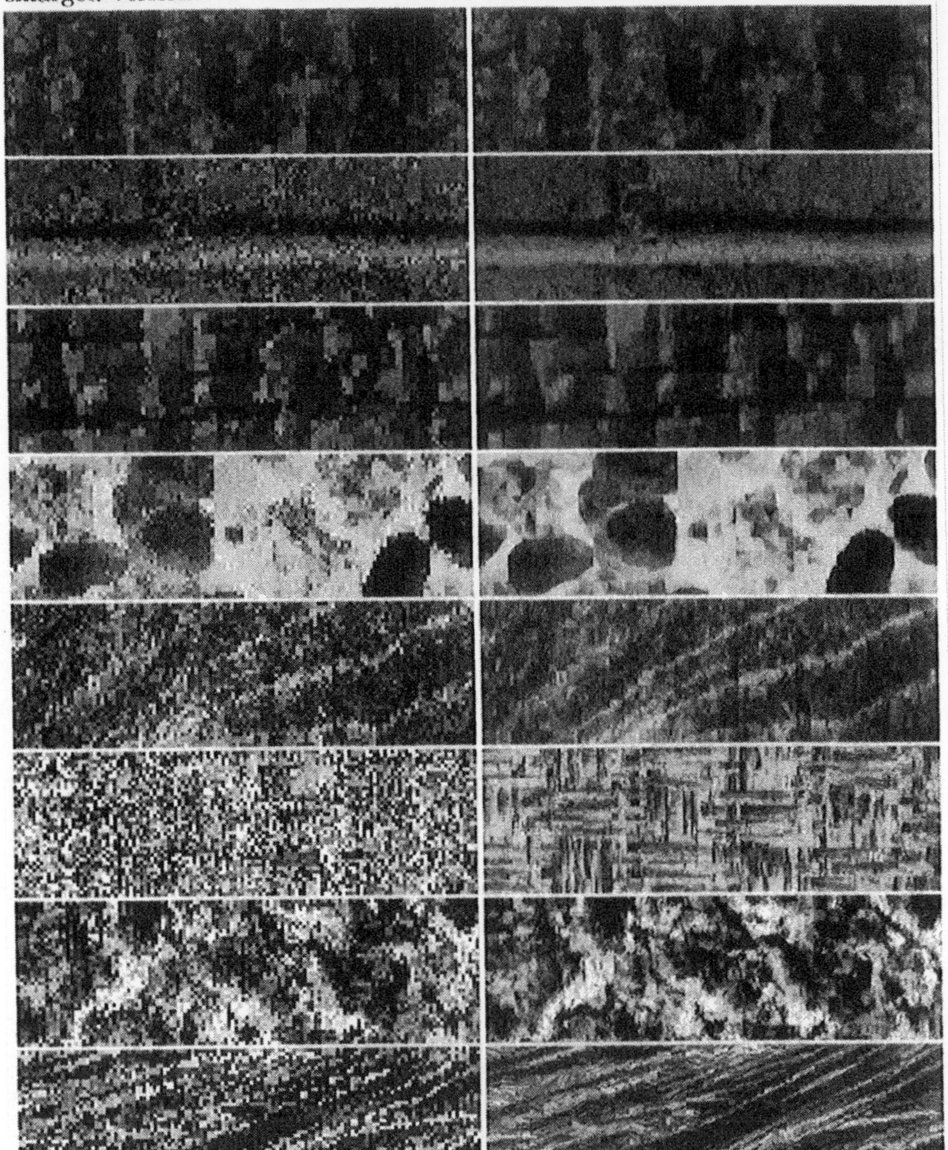

The two images for each texture correspond to the third and tenth iteration, respectfully. This shows that after three iterations, at high (scaled by a factor of 8 in both horizontal and vertical directions) resolutions, the texture image is recognisable. Then after a few more iterations the synthesised pixels are shown to be quite suitable in describing the texture type.

The following two image shows four single point perspective projected textures at different compression ratios and number of iterations for the original bark texture map. The left image uses approximately 15000 transforms displaying the first, second, fourth and eigth iteration in a counter-clockwise manner starting at the bottom. The right image shows eight iterations with approximately 1000 transforms, 2000 transforms, 4000 transforms and 8000 transforms.

Next the following two image shows similar stages of PIFS texture mapping for the brick texture.

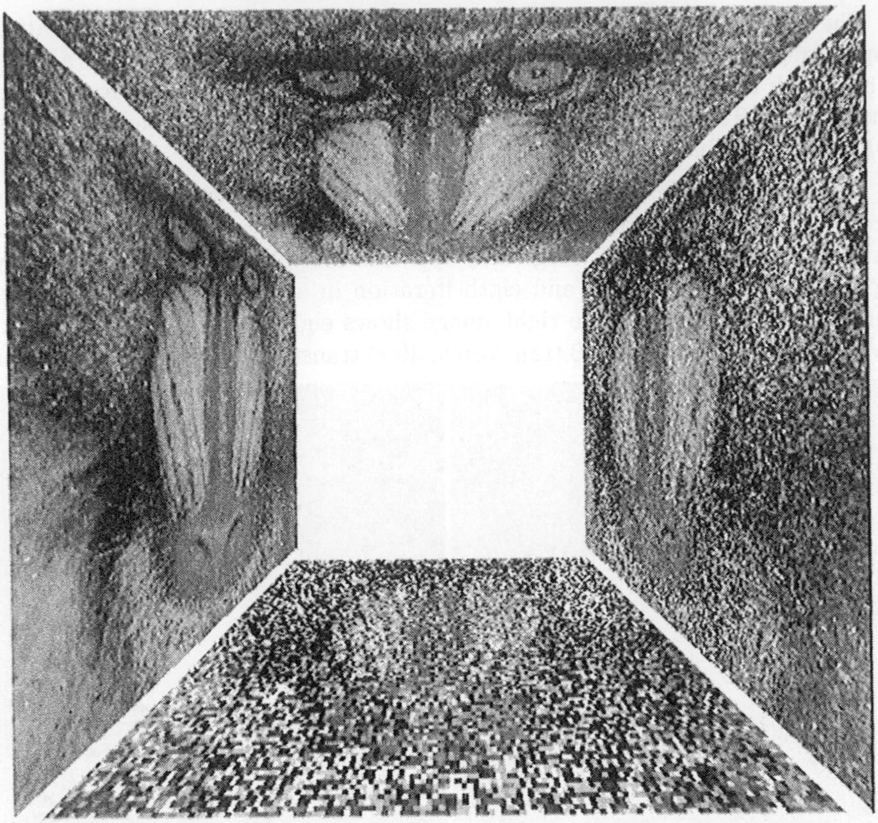

The left images show how the noise and other artifacts are progressively removed as the fixed point attractor image for the PIFS image is reached. The right images show how the fractal compression artifacts, including high frequency suppression and blocking effect the visual aspect of the perspective projection. Overlapping range blocks have been proposed and used to try and eliminate these edge effects. Quality is gained with a corresponding reduction in compression ratio. For texture mapping as compression ratio is potentially a secondary priority, this could be usefully applied.

In an annimated environment it is postulated that the previous display memory will be a close approximation to the final fixed point attractor and therefore can be used as the initial starting 'guess' image. This will hopefully increase the rate of convergence, and thus reduce the level of noise artifacts seen.

The final two images show the Mandrill image with similar stages of PIFS texture mapping as the two previous Brodatz textures. It is to be noted that the Mandrill image has a thin white border on the top and bottom to give it a square resolution, (512 × 512) from its original size of (512 × 480), without distorting the image. At high resolution there are a few artifacts which are due

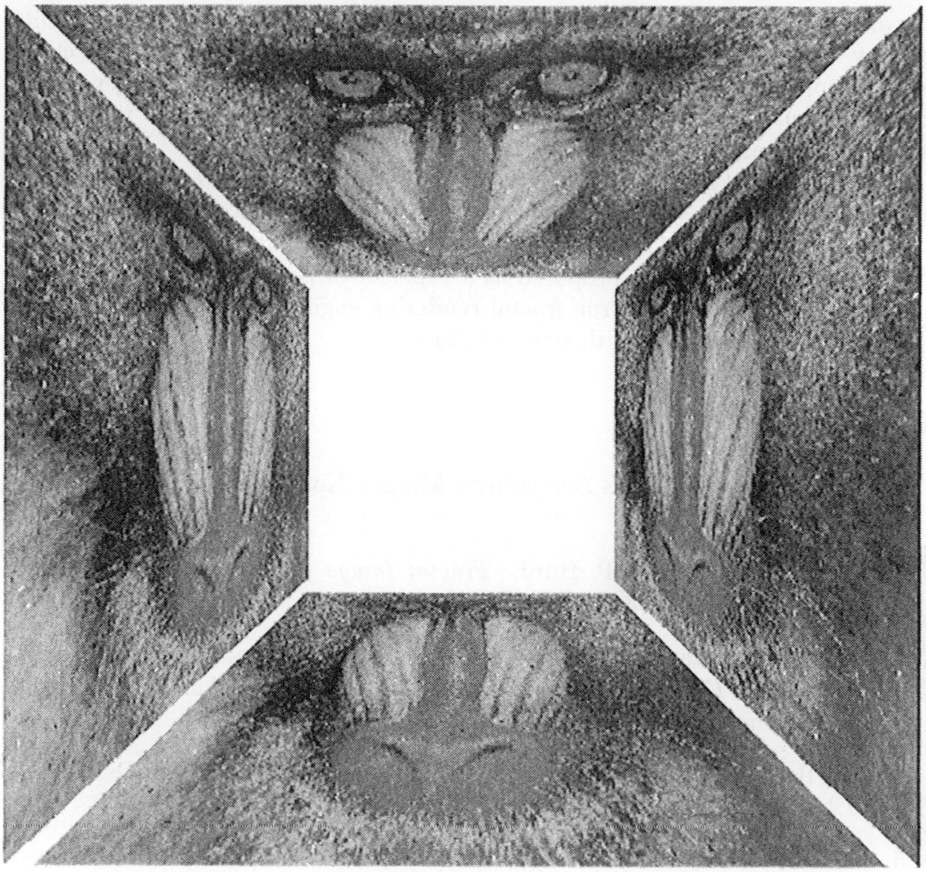

to the small number of sampling points, and not using any pixel anti-aliasing methods. The quality is still visually high and the process is fast enough for animation and full resolution enhancement.

5 Conclusions

This study and previous work [23] has shown the practical nature for exploiting compression artifacts, in this case to create more natural looking images. The issues mentioned need to be addressed, but the more important consideration is to address the visual acceptability of the imagery. Informal discussions with graduate students showed a marked divide with design students preferring the textual nature of the fractal compression, and computer science students preferring the pixellated enlargements.

In a recent panel discussion at Eurographics UK Chapter Conference 1999, Swansea [11], there were various views from the graphics supercomputing industry and modern games designers. It was proposed that even with the availability of very large and fast texture mapping memory, with the availability of cheap and fast processing engines, now in modern games consoles, texture generation and manipulation will become more functionally generated rather than using sophisticated mip-mapping tools.

Rendering using pure IFS definitions has also been proposed and implemented elsewhere [17]. This system has been adapted to include Gouraud shading and shadowing with extensions for Phong shading and anti-aliasing operations. As fractal geometry has been proposed as a superset of Euclidean geometry and it could be suggested that a true fractal rendering engine will be universally able to construct images with both types of object.

Bibliography

1. M.F. Barnsley. *Fractals Everywhere.* Morgan Kaufmann, 3rd edition, 2000. ISBN: 0120790696.

2. M.F. Barnsley and L.P. Hurd. *Fractal Image Compression.* AK Peters, Wellesley, Mass, 1993. ISBN: 1568810008.

3. Geoffrey Brindle. Anti-aliased textured using shape-lookup tables. In *Eurographics UK Chapter*, pages 87–94, April 2001. 19th Annual Conference held at University College, London.

4. Phil Brodatz. *Textures: A Photographic Album for Artists and Designers – 112 plates.* Dover Publications, Inc., New York, 1966.

5. P.J. Burt. The pyramid as a structure for efficient computation. In A. Rosenfeld, editor, *Multiresolution Image Processing and Analysis.* Springer-Verlag, 1984.

6. R.J. Cant and P.A. Shrubsole. Texture potential mip mapping, a new high-quality texture antialiasing algorithm. *ACM Transactions on Graphics*, 19(3):164–184, 2000.

7. Neil A. Dodgson. Image resampling. Technical report, University of Cambridge, Computer Laboratory, New Museums Site, Pembroke Street, CB2 3QG. United Kingdom, August 1992. TR-261.

8. David S. Ebert, F. Kenton Musgrave, Darwyn Peachey, Ken Perlin, and Steven Worley. *Texturing and Modeling.* AP Professional, Second Edition, 1998. ISBN: 0122287304.

9. Ahmet M. Eskicioglu and Paul S. Fisher. A survey of quality measures for gray scale image compression. In James C. Tilton, editor, *Space and Earth Science Data Compression Workshop*, pages 49–61. NASA Conference Publication 3191, April 1993.

10. Ahmet M. Eskicioglu, Paul S. Fisher, and Siyuan Chen. Image quality measures and their performance. In James C. Tilton, editor, *Space and Earth Science Data Compression Workshop*, pages 55–67. NASA Conference Publication 3255, April 1994.

11. Eurographics UK Chapter. *Conference Proceedings*, PO Box 38, Abingdon, Oxon OX14 1PX, 4-6 April 2000. 18th Annual Conference, Held at the University of Wales, Swansea ISBN 0952109794.

12. Yuval Fisher, editor. *Fractal Image Compression: Theory and Application to Digital Images.* Springer Verlag, New York, 1995.

13. James Foley, Andries van Dam, Steven Feiner, and John Hughes. *Computer Graphics: Principles and Practice.* Addison Wesley, Reading, Mass., second edition, 1996. The Systems Programming Series. ISBN: 0201848406.

14. R. Hamzaoui and D. Saupe. Rate-distortion based fractal image compression. In *Image Processing II.* Second IMA Conference on Image Processing: Mathematical Methods, Algorithms and Applications. 22-25 September 1998, Horwood, 2000.

15. Bill Jacobs, Yuvel Fisher, and Roger Boss. Image compression: A study of the iterated transform method. *Signal Processing*, 29:251 – 263, 1992.

16. Arnaud E. Jacquin. Fractal image coding: A review. *Proc. of the IEEE*, 81(10):1451–1465, October 1993.

17. Huw Jones. Exact object rendering using iterated function systems. In *Eurographics UK Chapter*, pages 27–36, April 2001. 19th Annual Conference held at University College, London.

18. John Kominek. Convergence of fractal encoded images. In *Data Compression Conference*, pages 242–251. IEEE Computer Society, March 1995. ISBN 0-8186-7012-6.

19. Ning Lu. *Fractal Imaging.* Academic Press, 1997. ISBN: 0-12-458010-6.

20. Peter R. Massopust. *Fractal Functions, Fractal Surfaces and Wavelets.* Academic Press, 1994. ISBN: 0-12-478840-8.

21. NASA – Space and Earth Science Data Compression Workshop, 2 April 1993. Discssion topic during morning session.

22. Matthias Ruhl and Hannes Hartenstein. Optimal fractal coding is NP-hard. In *Data Compression Conference*, pages 261–270. IEEE Computer Society, March 1997. ISBN 0-8186-7761-9.

23. Martin J. Turner. Design of fast fractal texture mapping. In *Eurographics UK Chapter. 17th Conference*, 13-15th April 1999. ISBN 0-9521097-8-6.

24. Martin J. Turner, Jonathan M. Blackledge, and Patrick R. Andrews. *Fractal Geometry in Digital Imaging.* Academic Press (Harcourt Brace and Company), 24-28 Oval Road, London NW1 7DX, June 1998. ISBN: 0-12-703970-8.

25. Alan Watt. *3D Computer Graphics.* Longman Higher Education, 3rd edition, November 1999. ISBN: 0201398559.

26. Lance Williams. Pyramidal parametrics. In *ACM Computer Graphics (SIGGRAPH '83 Proceedings)*, volume 17 (3), pages 1–11, July 1983.

Fractal Time and Nested Detectors

Susie Vrobel

The Institute for Fractal Research, Schachtenstr. 5, 34130 Kassel, Germany

Abstract

Fractal structures may be detected and described either by means of a non-nested detector/observer or by a nested one. A non-nested detector/observer registers the fractal structure on one level of description and thus detects inherent correlations only on that level. A nested detector, however, registers a fractal signal (in the case of temporal fractals) on several nested levels of description. It may then compare these levels, detect scale invariances, and anticipate the next nesting. An algorithm is introduced by means of which this nested detector anticipates scaling structures. Nested detectors with this anticipatory reading capability will register spacetime as a function of the scaling structure which recurs on all levels of description and reveal a scaling relativity.

1 Introduction: Levels of Description and Tangled Hierarchies

Whenever we attempt to describe aspects of the world such as the dynamics of a complex system, we necessarily limit our observation to certain abstractions of that system. This makes sense if we wish to focus on characteristics which we suspect play an important role in the dynamics of the system: We choose our target characteristics by selecting variables which best describe these characteristics. Our observation may also be limited by factors such as a limited range of our measuring apparatus or the amount of data we may process within a reasonable period of time. All these constraints generate abstractions which I denote as "levels of description" (LOD) after D.R. Hofstadter's concept of level differentiation and interaction [3]. These levels of description may differentiate between different levels of complexity, some of which differ in quantity, others in quality. An example of LODs defined by different qualities is Escher's etching *Drawing Hands*, which depicts a left and a right hand, each holding a pencil. With the respective pencils, the right hand is drawing the left one while the left is drawing the right one. This tangled hierarchy ("strange loop")[1] is visible

[1] According to Hofstadter, strange loops or tangled hierarchies are present, "if ,... something *in* the system jumps out and acts *on* the system as if it were *outside* the system." Hofstadter, Douglas R.: *Gödel, Escher, Bach - An Eternal Golden Braid.* Vintage Books, N.Y. 1980, p. 691.

only on the level of the etching itself. The tangled hierarchy can be untangled by taking a step back, i.e. by taking account of another level of description. In the case of the *Drawing Hands*, the level of decription which untangles the strange loop is that of the artist, M.C. Escher, who, by drawing both hands, restores a perfectly untangled hierarchical order. Escher is the inviolate level, which cannot be accessed from the tangled level, let alone be modified in any way. The inviolate level is a constraint to entanglement. In fact, its existence can be suspected from the tangled hierarchy level only as a result of the fact that a strange loop exists. According to Hofstadter, below every tangled hierarchy lies an inviolate level which disentangles the strange loop.

The inviolate level is visible only to an observer who is located outside the tangled system. From an internal perspective, nothing but the tangled hierarchy is visible and the existence of the inviolate level may only be deduced logically.

Different levels of complexity may lead to the same effect: causative loops can be described as a result of a "translation causality". Depending on the observer's perspective, a phenomenon may be described as being the result of either or both upward and downward causation. Upward causation corresponds to the reductionistic explanation, where entities on lower LODs cause phenomena on higher LODs (e.g. photons forming a lightbeam), downward causation describes processes on lower LODs as being constrained by organization principles of higher LODs (e.g. a laser, where photons on a lower LOD are constrained by organizing patterns on a higher-level LOD).

Upward or downward causation is, in Hofstadter's terminology, a form of "level-crossing": a translation between LODs. Such translations between levels may appear as causation: Hofstadter points out that an explanation of a phenomenon is often a description of the same phenomenon on a different LOD: "Moreover, we will have to admit various types of 'causality': ways in which an event at one level of description can 'cause' events at other levels to happen. Sometimes event A will be said to 'cause' event B simply for the reason that the one is a translation, on another level of description, of the other" [4]. The concept of condensation introduced later in this paper is a variation of this notion of level-crossing.

2 Fractals

Fractals are many-leveled phenomena which exhibit structure on various LODs. Their LODs vary in the degree of complexity, as these first iterations of the Koch Curve in Figure 1 shows.

There are, however, also differences in quality between the LODs of a fractal, which differentiate between tangled or crossable levels and the inviolate level. It is this fractal analogue to the inviolate level which my Theory of Fractal Time is concerned with. The concept of the Prime, which is introduced below, forms the inviolate level for nested scaling structures.

One way of defining a fractal is to require a many-leveled structure. The

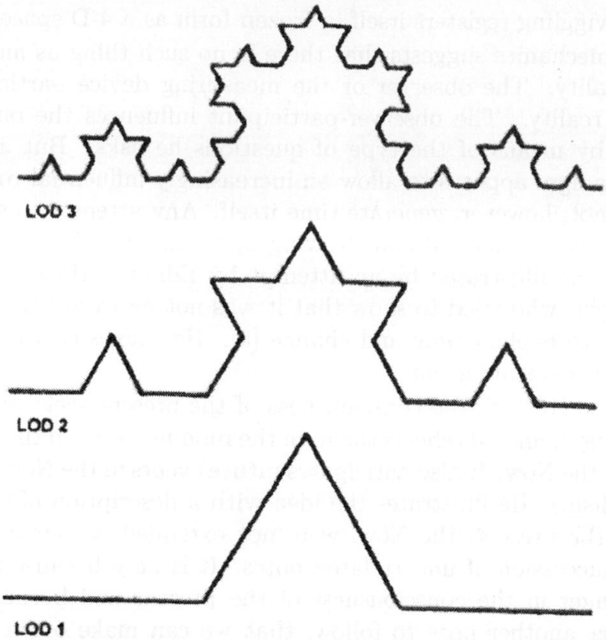

Figure 1. The first iterations of the Koch Curve

many-leveledness does not necessarily need to be inherent in the structure we observe or measure. It may also be the many-leveled way of observing or measuring a system which generates a fractal structure. Barnsley's Box Counting Method is such a fractal way of analysing spatial and temporal structures [1]. When we observe fractal time series, we should keep in mind that, depending on where the interfacial cut is made, a fractal on the interface may be a manifestation of the intrinsic dynamics of the system itself, of the measuring chain, or of the observer. If we wish to talk about and analyse fractal time series, we must first define a suitable concept of time on which we may base our method of analysis.

3 Concepts of Time

Replacing Newtonian absolute time by Einstein's relative concept of time, which allows for relative observer frames, did not change the fact that time was still regarded as simply "being there", i.e. existing independent of an observer or detector in a pre-established framework. This is, as Wheeler put it, "a view of nature in which every event, past, present or future, occupies its preordained position in a grand catalog called spacetime" [16]. He points out that "spacetime does not wiggle. It is 3-D space geometry that undergoes the agitation. The history of its wiggling registers itself in frozen form as a 4-D spacetime". [17].

Quantum mechanics suggests that there is no such thing as an independent observer of reality. The observer or the measuring device participates in the generation of reality. The observer-participant influences the outcome of his measurement by means of the type of questions he asks. But although each successive paradigm appears to allow an increasingly influential role for the observer, he cannot, however, generate time itself. Any attempt to show that the subject can generate time without drawing upon an objective time is bound to fail. This may be illustrated by an attempt by Edmund Husserl, the German phenomenologist, who tried to show that it was not necessary to draw upon an objective time to explain time and change [5]. He suggested the subject itself should be the generating agent.

According to Husserl, the consciousness of the present seeks out past events by remembering them and reflects them, in the modified way, in the consciousness of the present, the Now. It also anticipates future events in the Now (which allows for intentionalism). He illustrates the idea with a description of the perception of a tune. If the present, the Now, were not extended, we could never hear a tune, only a succession of uncorrelated notes. It is only because the notes just played still linger in the consciousness of the present and because of the fact that we expect another note to follow, that we can make out a tune. If this process is continued, a nested and nesting structure appears with a cascade of remembered, or still lingering, notes which are reflected in the now. Such a concept of the Now must exhibit extension, in order to host both the protention and retention (anticipating and remembering). Husserl assumed that the subject generates time by nesting events in an extended Now. This claim that the subject generates time in this way is a fallacy, however, because the notion of the consciousness of the present, the Now, already assumes a succession, which must take place against the background of frame time: Another level of description must be assumed as a frame time in which these successions can take place. This frame time cannot be observer-generated, as the observer is himself embedded in this frame time. Therefore, we must assume that time is real and exists independent of the observer. The observer cannot generate time. As an observer-participant, he may, however, be in a position to structure it. The notion of a nested structurable time within an extended Now provides a suitable concept for fractal time series analyses, as it is based on many-leveled nested structures into which the observer may zoom (thereby extending previously point-like Nows to ever more structured intervals).

4 A Theory of Fractal Time: $\Delta t_{\text{depth}}, \Delta t_{\text{length}}$ and $\Delta t_{\text{density}}$

I have drawn upon Husserl's notion of a nested time structure and an extended Now to formulate my Theory of Fractal Time [13]. The notion of Fractal Time allows the description of a process independent of a specific, arbitrarily chosen level of description. In contrast to the time of physics, t, the fractal time concept differentiates between

- Δt_{depth}, the depth of time, which is defined as the number of nestings of a recurring structure (i.e. the number of LODs);
- Δt_{length}, the length of time, which is defined as the number of incompatible events on one LOD; and
- $\Delta t_{\text{density}}$, which is the fractal dimension, i.e. the relation between Δt_{length} and the scaling factor. (This concept, which - loosely speaking - describes the density of time, is not needed to describe the main concepts of the theory, such as the Prime and condensation. Therefore, I shall not extrapolate on $\Delta t_{\text{density}}$, but focus on the notions of Δt_{depth} and Δt_{length}).

Whenever we wish to specify the length of an interval Δt, we have to define the units in which this length is to be measured. This is only possible if we first specify the LOD which is to serve as a reference for the LOD considered. After all, *measuring means comparing LODs.* Therefore, Δt_{depth} logically precedes Δt_{length}, and the differentiation between Δt_{depth} and Δt_{length} is an epistemological necessity.

Based on this concept of Fractal Time, one may construct a fractal clock. Imagine tiny pointers following the circumference of the Triadic Koch Curve (I use the Koch curve for didactic purposes only. As a mathematical artifact, it does, of course, not correspond to any underlying physical process.) in Figure 2a in such a way that they tick simultaneously from corner point to corner point, on all LODs, as lined out in Figure 2b. The first pointer ticks three times to complete the circumference, while the second pointer ticks, simultaneously, 12 times while the third pointer ticks 48 times and so on ad infinitum.

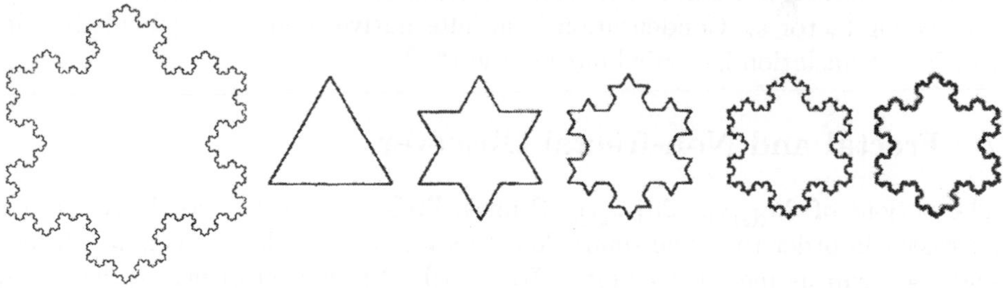

Figure 2. a) The Triadic Koch Curve, b) Pointers ticking simultaneously on all LODs.

5 Primes and Temporal Natural Constraints (TNCs)

The Triadic Koch curve is an idealized structure. When look at real data, i.e. structures in Nature, however, there is always an upper and a lower limit to this type of scaling behaviour. The upper limit of a self-similar domain I have denoted as the Prime.[13] The Prime can be defined as the structure at the "top end" of a nesting cascade, which is itself part of and nested in the "next-lower" structure, but has no nesting potential itself. The Prime has, as the last nested structure, a limiting effect on the entire structure's nesting potential: it is a nesting constraint. This also implies that the Prime has a limiting effect on the extention of Δt_{depth}, which makes it a Temporal Natural Constraint (TNC) [14].

6 The Prime Structure Constant (PSC) and Condensation: A Different Set of Level-Crossing

In a self-similar nesting cascade, the structure of the Prime recurs on all LODs. This recuring structure serves as a reference scale between the various LODs and allows the description of internal correlations with respect to this structure. The lengths of the intervals ($\Delta t_{\text{tlength}}$) covered by this structure, which I have denoted the Prime Structure Constant (PSC) [14], vary from LOD to LOD. However, the PSC may be made congruent on all LODs by means of an appropriate variation of scale on those levels. If the PSC is set as a constant, a new kind of relativity results, in which time is distorted with respect to the PSC.

The state of congruence on all LODs I have denoted as *condensation* [13]. Condensation is a property of spacetime, generated by congruent scale-invariant nestings (cf. Figure 3).

Condensation may be quantified as condensation velocity $v(c)$ and condensation acceleration $v(a)$.The quantities required for the determination of $v(c)$ and $v(a)$ are Δt_{depth} and Δt_{length}. In a self-similar structure, the quotient of Δt_{length} of LOD 1 and Δt_{length} of LOD 2 equals the condensation velocity $v(c)$ for LOD 2 nested in LOD 1. For scale-invariant structures, $v(c)$ is identical with the scaling factor s. Condensation is an alternative mode of level-crossing, in which all translation is carried out via the PSC.

7 Fractal and Non-fractal Observers

The notions of Δt_{depth}, Δt_{length}, Primes, PSCs, TNCs and condensation are necessary in order to define smart detectors which are able to recognize Primes and use them as measuring units. To describe the notion of nested detectors, a further differentiation must be made between the concepts of a fractal and a non-fractal observer.

If an observer is to describe a process, two modes of observation, or two types of observers, must be differentiated: the fractal and non-fractal observer. The idea of a fractal observer draws upon Rössler's question "If I were the only chaotic system in the universe, would not the whole universe appear chaotic to me?" [7] and his idea "... that a universe that is chaotic itself ceases to be chaotic as soon as it is observed by an observer who is chaotic himself" [8]. If we

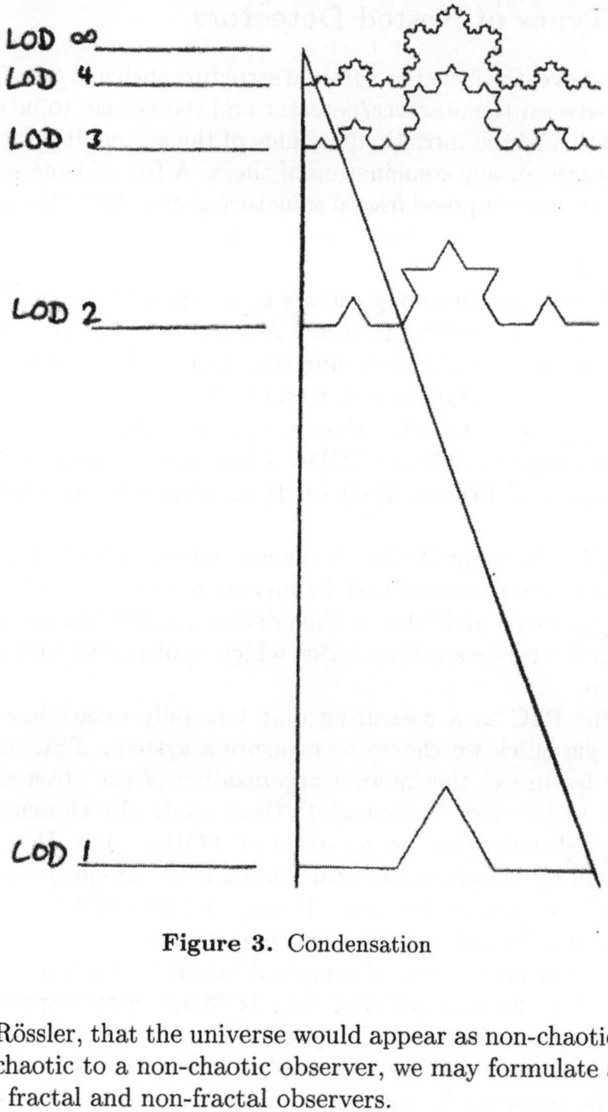

Figure 3. Condensation

assume, with Rössler, that the universe would appear as non-chaotic to a chaotic observer and chaotic to a non-chaotic observer, we may formulate an analogous hypothesis for fractal and non-fractal observers.

A non-fractal observer perceives reality by registering individual LODs separately, from an unnested perspective. To him, the Prime structure covers intervals of varying Δt_{length} on the individual LODs, which results in the belief that the observed Prime structure is not congruent on all LODs, but distorted, i.e. condensed or dilated with respect to an "absolute" frame time. A fractal observer perceives reality by registering individual LODs simultaneously superimposed, from a nested perspective which may induce condensation. To the fractal observer, the Prime structure covers a constant Δt_{length} on all LODs, a perspective which results in the belief that time is distorted with respect to the PSC.

8 Three Types of Nested Detectors

As mentioned above, the fact that a fractal structure such as $1/f$ noise appears on the interface between the observer/detector and the system to be observed may be a manifestation of the intrinsic dynamics of the system itself, the measuring chain, the observer, or any combination of these. A fractal time series may thus be the result of a superimposed fractal structure generated by the internal nested organization of the detector. Such a nested detector could come in three types A, B and C [15]:

Detector A registers incoming signals in a vertically uncorrelated way, i.e. it does not compare its LODs. Detector A registers vertically uncorrelated sequences and can thus not detect scaling structures. (This detector may as well be substituted by several non-nested detectors.)

Detector B registers incoming signals in a vertically correlated way, i.e. it compares its readings on different LODs. Therefore, it is capable of recognizing scaling structures and Primes. Detector B registers a fractal structure such as $1/f$ noise.

Detector C is a Detector B which has been trained or has learned to recognize Primes and turn the structure of each Prime into a constant (the PSC) in order to generate congruence on all LODs. It then defines the PSC as the next measuring unit. Detector C registers a time series which is distorted with respect to the Prime structure.

Defining the PSC as a measuring unit hopefully makes less arbitrary the choice of the yardstick we choose to measure a system. PSC units pick up a characteristic feature of the internal organization of the observed system and may turn out to be more "meaningful" than arbitrarily chosen units, as they reflect the nested underlying processes on all LODs. Thus, Detector C follows a clue presented by ubiquitous natural scaling laws. Ubiquity should make any observer highly suspicious, however. It may indicate that it is not the world around us which is fractal, but the way we look at it.

Detector C generates a fractal temporal interface which reveals a new type of relativity, where we may say that time is "bent" with respect to the Prime structure. Everything Detector C registers, is, of course an internal measurement. How to communicate the outcome of an internal measurement is a problem which might be overcome by using the Prime as a universal translator between different Detectors C. A Detector C acts as a condensator which may be generated by teaching a Detector B where to look.

9 An algorithm for generating a condensator (Detector C) by teaching Detector B where to look

Step 1: Read incoming signals V simultaneously on LOD_n, LOD_{n+1} and LOD_{n+2}. Register and remember the structure of the incoming signals V.

Step 2: Compare LOD_n with LOD_{n+1} until a self-similar nesting is detected.

Step 3: If a self-similar nesting is detected, search for self-similar nesting on the next level LOD_n, LOD_{n+1} and LOD_{n+2} on the basis of the scaling factor s relating LOD_n and LOD_{n+1}. Continue for a pre-defined number of steps (LODs).

Step 4: Set the change in variable V over an interval t which recurs on all LODs as the Prime structure.

Step 5: Generate successive filters which mimick the Prime structure on all LODs.

Step 6: Generate a fractal interface consisting of these successive filters which filter out everything except the Prime structure on all LODs.

Step 7: Set the Prime structure as a constant so that $\Delta t_n = \Delta t_{n+1} = \Delta t_{n+2}$ (for a pre-defined number of steps (LODs)).

Step 8: Carry out simultaneous readings with the PSC as the new measuring unit.

Step 9: The fractal interface registers a non-fractal structure. It acts as a condensator for which $\Delta t_n = \Delta t_{n+1} = \Delta t_{n+2}$ for a pre-defined number of steps (LODs). "Display" readings. (Detector C is an endo-detector/observer whose readings cannot be communicated to an exo-detector/observer. It can, however, communicate with other equivalent endo-observers via the Prime).

One nesting (2 LODs) is sufficient; a deeper nesting cascade is preferable. The scaling factor may vary: In this case, short self-similar domains with constant s (with $c(v)$ = constant) may be followed by $c(a) \neq 1$ and vice-versa.

During the interface-forming phase, events are not registered in real time, as the detector must first register the signals and compare them by means of a program in order to make out a self-similar pattern for the first LODs. Once the interface is formed, however, the registering of incoming signals is in real time with the PSC as the new measuring unit. (This unit of time is indivisible as it has no nesting potential (so any further differentiation would not make sense) and forms an extended Now in Husserl's sense).

What, then, does this detector register? As to a fractal observer, a fractal would not appear as a fractal to that detector/observer, but as a non-fractal structure which is the Prime.

This methods works best for data which exhibits discrete, exact self-similarity. Real data, however, tends to display statistical self-similarity, i.e. the PSC may be the change in a variable V over an interval t. In order to work with real data, statistical self-similarity of nestings x and y must be defined, on a different LOD, as discrete self-similarity for nesting z.

Step 9 produces an internal reading which cannot be observed directly from outside this detector. There are two possible scenarios which would allow an observer to access this internal reading. In the first scenario, the observer *is* Detector C. This would not help, however, if the observer wished to communicate his findings to another observer.

The second option is to use the Prime as a universal translator.

10 The Prime as a Translator between Endo-Perspectives

Different observers/detectors are likely to have different LODs. Therefore, intervals which are "covered" by the PSC vary in (tlength for different observers. The Prime itself, however, is a non-structurable entity, a unit of time, which connects and relates all private (endo) LODs. Primes set, as temporal natural constraints (TNCs), a limit to the "structurability" of time by the observer. They are invariant entities of time in a covariant world and distortion limits in the sense of Rössler's interfaciology: "The very notion of limits implies the adoption of two perspectives simultaneously. (Second,) any distortion limit 'splits' the single (exo) reality into many different internally vàlid (endo) realities."[11]

The Prime is such a limit: it suggests the existence of other simultaneous perspectives, other observer frames/detectors, from whose perspectives the Prime is also a non-structurable entity. If all observers/detectors measure in PSCs, they all use the same yardstick, although extensions in Δt_{length} may vary, and are therefore translatable into each other. The Prime may act as a translator with custom-made translation procedures, linking otherwise incompatible endo-realities. In fractal structures, Primes form the inviolate level.

The inviolate level is, by definition, accessible only from outside the observed systems. There is a way out, however: If the the tangled LODs and the inviolate level share the same basic recuring structure, as in a fractal nesting cascade, the self-similar structures on all LODs make the inviolate level accessible via the tangled LOD.

What makes an LOD a tangled level is the fact that the observer's window to the world, the Now, is inherently subjected to a strange loop: Whenever we try to define what we mean by time (or a unit of time), we have to do this while we already are embedded in what we try to define: time (or a unit of time). This logical loop suggests the existence of an inviolate level and makes the observer an observer-participant who is positioned on a tangled level, allowing only an endo-perspective. The exo-perspective is, according to Rössler, reserved for a superobserver.[10]

11 Nested (Fractal) Detectors/Observers are Created by Generating LODs

Only a nested observer/detector can make out Primes and use the PSC as a measuring unit on all LODs in order to induce condensation. At least two simultaneous nested detectors of type C are required to register differing endo-realities of a single exo-reality, i.e. the Prime. The Prime is governed and generated by processes of the exo-world (this is compatible with the necessary assumption that time exists independent of an observer).

In order to generate differing LODs, detectors of type C must be generated in different environments to ensure they "learn" under differing conditions. The detectors generate differing fractal interfaces, which project the PSCs onto their private LODs.

The way a detector learns, i.e. generates LODs, varies considerably, depending on the type of detector considered. The incoming signal could have a reinforcing effect on those parts of the observer/detector it hits. With every registered impact, the detector becomes more and more sensitive to incoming signals.

An artificial intelligence may, for example, imitate the way a child learns spatial and temporal expressions by generating new LODs: Clark's complexity hypothesis states that, before learning spatial and temporal expressions, a child must first master so-called of rules of application, which presuppose attributes such as fields of perception and direction. Clark's complexity hypothesis, is based on the correlation between human levels of perception and the appropriate language levels [2]. It states that the order in which spatial concepts are acquired is forced by (the acquisition of) rules of application which include direction, point of reference and dimension:

"If, of two terms A and B, B requires all the rules of application A requires plus an additional one, then A is acquired before B. This idea is illustrated for the prepositions *in*, *into*, and *out of*: *in* presupposes a three-dimensional space; *into* presupposes a three-dimensional space and a positive direction[2]; *out of* presupposes a three-dimensional space, a positive direction and a negation of this direction. According to Clark's complexity hypothesis[3], these prepositions are acquired in the following order: First *in*, then *into*, and finally *out of*." [13]

Clarke's rules of application are one example of how LODs can be acquired by an intelligent observer-participant. Although the nature of the generated LODs strongly depends on the underlying system, the example of LOD-generation in language acquisition is representative for all level-generating systems as far as

[2]According to Clark, the term *positive* means, in this context, *in the stronger perceptual field.* Cf. Clark 1973.

[3]The *complexity hypothesis* makes the following further predictions:

1. In antonymous pairs, the positive term will be acquired before the negative one (e.g. *into* before *out of*);
2. *At*, *on* and *in* are acquired before *to*, *onto* and *into*, since the latter require, in addition, a direction;
3. Location prepositions such as *at*, *on* and *in* are acquired before correlative location prepositions such as *above* and *in front of*, since the latter require, in addition, a point of reference;
4. *Tall* and *short* will be acquired before *thick* and *thin*, since the latter require an additional dimension;
5. Unmarked terms will be acquired before marked terms. The positive term is acquired before the negative one and the positive term determines the dimension: *long* (+), *short* (−) ⇒ dimension: *length*.

the defining characteristics are concerned. These defining characteristics of LOD-generation are:

Mutual modification: The generation of each new LOD is the result of interaction of the observer/detector with his/its environment. Both incoming signals and the detector's "perspective" i.e. its way of filtering and registering this data, change the interface between observer/detector and the outside world.

Individuality: Each observer/detector generates his/its private nesting cascade of LODs. As a complex internal organization of the observer/detector results in intentionalism on the side of the observer/detector, it is unlikely that their LODs match. The different degrees of complexity in the environment also contribute to very individual LOD-generation. (In the case of language acquisition, a Chinese speaker and a German speaker would have formed very different interfaces.)

Constraints: For natural systems, the nesting of LODs is limited. This is true for both scaling systems and non-scaling systems. In scaling systems, the self-similar domain has a lower and an upper limit, which is the Prime. In non-scaling systems, the limitation manifests itself in the degree of complexity nested LODs can reach before the quantitative increase in complexity turns into a new quality. A new quality opens up a new domain which may or may not be self-similar when increasing complexity generates new LODs. TNCs such as PSCs are generated in this way. (In the case of Clarke's language acquisition example, a further rule of application would produce not a more complex preposition, but either a new concept other than a preposition or something which would not make sense at all. Either case implies a limitation on more or less complex prepositions.)

In general, hearing new expressions before one has mastered the required rules of application or registering undifferentiated incoming signals leave a trace in the observer/detector, which changes this measuring device itself. New experiences or new incoming signals then generate new LODs which, in turn, influence the subject's or the detector's interface. For an observer, this would modify his perception and empirical knowledge of time. It is also conceivable that, after a while, when a certain reinforcement stage is reached, the observer/detector ignores these frequently incoming signals (just like the brain does). In this case, the observer's/detector's interface would be moulded as a "negative form."

12 Bridging Endo- and Exo-Perspective

From the foregoing considerations we may assume that all the conventional non-nested observer/detector ever registers (unless it is a Detector of type C) are distortions generated by his/its LODs. These LODs act like a prism, unfolding a superimposed, undifferentiated incoming signal into a many-leveled phenomenon which we describe as a fractal.

Therefore, the conventional non-nested observer/detector cannot perceive/register a Prime, not even the PSC, as he/it sees the PSC distorted in time and superimposed, as a result of prism-like LODs. LODs are private generations, which create an individual interface and endo-perspective for the observer/detector. Primes, however, are universal: the inviolate level, which generates the exo-perspective, is made up of Primes.

An observer/detector of type C, however, may induce condensation and catch a glimpse of the Prime, i.e. the exo-perspective. We may then, by means of the PSC, compare the readings of different detectors of type C which have learned/evolved in different environments and have therefore generated different interfaces, so they have no "common denominator" other than the PSC.

The Prime is the only entity (of time) which is invariant under transformations such as time dilations or contractions. It forms the inviolate level of time and is the scale relativity analogue to c in General Relativity. See also[6].

Detector C's readings are an internal process, however, so how could an outside observer know whether or not condensation has occured and whether it has been able to access the Prime?

It is conceivable to create systems on the modelling level, as Rössler suggested [10], with ourselves as superobservers situated outside the system. We may then conduct the following experiment: Two detectors of type C which have learned/evolved in different environments and have thus generated different interfaces are confronted with the same question regarding the PSC. If both observers/detectors of type C have generated sufficient LODs which may then host the PSC, they may use this PSC as a measuring unit. This may induce condensation, which allows them to access the Prime, i.e. the exo-perspective.

If we now posed the same question concerning the distortion caused by the induced condensation to both endo-observers/detectors, both should give the same answer. If both gave the same answer, this would imply that they both recognized a PSC, although their LODs are not the same and they do not share any common denominators other than the PSC, that they use it as a measuring unit and that they induce condensation. We may then conclude that both endo-observers had access to the same exo-perspective via the Prime, as the Prime is simply a PSC without nesting potential.

What type of question would serve as a suitable candidate to be posed to the detectors? The question would have to produce an answer about distortions caused by condensation, which can only be answered as a no/yes-question, i.e. with a 0 or a 1, where the detector would either say "yes" by switching itself off or say "no" by continuing to register incoming signals.

One such question may be: "Is Δt_{length} dilated on all your LODs in such a way that the previously registered distortions of the PSC have disappeared and are now congruent?" If this were true for both detectors, they would decide that the answer is "yes" and switch themselves off.

More complex PSCs require more complex questions about the characteristic distortions in their wake. It is, of course, harder to define a PSC consisting of the property "preposition" than a PSC with the property "Koch Curve initiator".

Condensation is induced only by a nested scaling structure. Therefore, a mere nesting of different levels of complexity does not suffice.

On a philosophical note, the Prime as a universal translator may reconcile the Now of the observer with the time of physics, t.

If we assume with Rössler, that "Nowness (...) is pure interface" [9], and that the mere existence of the Now implies a tangled hierarchy and therefore reveals the existence of an exo-perspective, we may conclude the possible existence of other equivalent endo-observers.

The endo-perspective deals with physics from within, with the endo-world of an observer-participant. The exo-perspective is that of a superobserver, a demon who is not part of the system he observes/controls [10].

The Now belongs, as a subjective phenomenon, to the endo-world of an observer-participant. However, the subjective Now of modal time has no equivalent in the relational time of physics t, which belongs to the exo-world of the superobserver. Neither Kant's approach, nor that of McTaggart or Husserl, could bridge the gap between a time which exists independent of an observer and the subjective time of the observer, whose window to the world is limited to the Now [13].

The Prime may be able to bridge this gap, as it can be accessed from both the endo and the exo world and is therefore a qualified translator between the two perspectives. Primes are "common denominators" for different observers with different observer frames. They form an inviolate level which may not be modified itself but can be used as a universal translator between different observer frames. Therefore, Primes are promising candidates for contacts and translators between endo-perspectives.

13 Possible Applications

Nested detectors of type C, which are capable of inducing condensation, may be used as a universal translation device between different endo-perspectives. Such different endo-perspectives exist, wherever participants in interactive cyber-games interact. The programmer would take the role of the superobserver from the exo-perspective and define the Primes which "govern the game", as they provide natural constraints to the endo-participant's interaction.

A further application of nested detectors of type C lies in the field of memory formation. Memory is correlation. Therefore, correlations such as those generated by PSC allow a more differentiated account of memory/correlations, as they add to autocorrelations in Δt_{length} the vertical correlations of Δt_{depth} and, via the Prime of the exo-perspective, correlations between otherwise incompatible endo-observers.

14 Outlook

There is another conceivable way of detecting condensation (apart from simulations). If we imagine ourselves not as superobservers conducting a simulation, but as observer-participants within a simulation, we could come to the conclusion that an inviolate level, an exo-perspective exists, if we encounter a tangled hierarchy or strange loop on our own level: Below every tangled hierarchy, there lies an inviolate level. The observer's Now, the window through which he can access reality, provides this strange loop.

The endo-observer may access the inviolate level, which is formed by Primes, by inducing condensation. In order to achieve this, the observer would have to generate as many LODs as possible to create an optimal internal differentiation. The resulting interface would be capable of picking up Prime structures. It would, however, require a trained observer to notice that a condensation has taken place to begin with. Such an event may be ignored by an observer as the result of a selection effect [12,14]: We only notice structures which are meaningful within our own endo-perspective. This constraint is a prerequisite which allows us to make sense of the world and create a non-counterfactual reality. We have learned to forget where to look for condensation scenarios.

There may be an way, even for well-adapted brains, to detect condensation: As observer-participants, we may look out for inadvertently induced condensation, which would manifest itself in spontaneously disappearing temporal distortions. This means that the observer suddenly recognizes that all he had been able to see so far was a distortion of the non-distorted, untangled structure he would then be able to see. It would take the disappearance of this distortion to make him aware of its existence and open up a new pespective. It is a situation reminicient of the situation in Plato's cave. Such "glimpses of the exo-world" are, however, by definition, a highly subjective phenomeneon and therefore not easily communicable. It may be appreciated as an expansion of one's private event horizon, but for a scientifically sound approach, we would have to confine ourselves to interpreting Detector C's readings.

Bibliography

1. Barnsley, Michael, *Fractals Everywhere.* Academic Press, London 1988.

2. Clark, H.H., "Time, Space, Semantics and the Child", in: T.E. Moore (Ed.), *Cognitive Development and the Acquisition of Language.* Academic Press, London 1973.

3. Hofstadter, Douglas R., *Gödel, Escher, Bach - An Eternal Golden Braid.* Vintage Books, New York 1980. Hofstadter, 1980, p. 709.

4. Husserl, Edmund, *Vorlesungen zur Phänomenologie des inneren Zeitbewußtseins.* (1928) Niemeyer, Tübingen 1980.

5. Nottale, L., "Scale Relativity, Fractal Space-Time and Quantum Mechanics", in: M.S. Naschie/O.E. Rössler/I. Prigogine, *Quantum Mechanics, Diffusion and Chaotic Fractals.* Pergamon. Elsevier Science, Oxford 1995, pp. 105-111.

Before explaining the stretch-and-fold transformation and its cryptographic application, we first identify the relationship between the two branches of mathematics — chaos theory and cryptography. Mathematical or deterministic chaos is a dynamic system, characterized with a 'complex' and 'unpredictable' behavior. Intuitively, this property suits the requirements of a digital encryption system — on the one hand computer-based cryptosystems are deterministic; on the other, they must be cryptographically unpredictable. Practically, the last property implies that given certain information on the ciphertext and the plaintext, a cryptanalyst should not be able to predict the cryptographic transformation and recover the key or the message.

There are several sufficient conditions satisfied by a dynamic system to guarantee chaos [5]; the sensitivity to initial conditions and topological transitivity are the most common. Formally, a chaotic continuous-state discrete-time system (continuous chaos for short) is a dynamic system $\mathcal{S} = \langle X, f \rangle$ with two properties:

1. Given a metric space $X \subseteq R^N$ and a mapping $f : X \to X$, we say that f is *topological transitive* on X if for any two open sets $U, V \subset X$, there is $k \geq 0$ such that $f^k(U) \cap V \neq \emptyset$.
2. The map f is said to be *sensitive to initial conditions* if there is $\delta > 0, k \geq 0$ such that for any $x \in X$ and for any neighborhood H_x of x there is $y \in H_x$, such that $\left|f^k(x) - f^k(y)\right| > \delta$.

The most distinctive characteristic of a chaotic system is bifurcation, i.e. the divergence of trajectories from adjacent starting points. A common measure of bifurcation is the Lyapunov exponent, denoted by λ and defined for continuous-state discrete-time systems by

$$|f^n(x_0 + \varepsilon) - f^n(x_0)| = \varepsilon e^{n\lambda},$$

where x_0 is the initial condition, f^n is a chaotic map applied n times and ε is a small perturbation. The formal definition of the Liapunov exponent is the limit of λ for $\varepsilon \to 0$ and $n \to \infty$ given by

$$\lambda(x_0) = \lim_{n\to\infty} \lim_{\varepsilon\to 0} \frac{1}{n} \log \left| \frac{f^n(x_0 + \varepsilon) - f^n(x_0)}{\varepsilon} \right|$$

or

$$\lambda(x_0) = \lim_{n\to\infty} \frac{1}{n} \sum_{k=1}^{n} \log |f'(x_k)| = \lim_{n\to\infty} \frac{1}{n} \log \prod_{k=1}^{n} |f'(x_k)| \tag{1.1}$$

For each k, $f'(x_k)$ tells us how much the function f is changing with respect to its argument at the point x_k. This derivative expresses the magnitude of change in the transition from x_k to x_{k+1}. The limit of the average of the derivative logarithms over n iterations is taken to provide a measure of how fast the orbit changes as (discrete) time propagates. A positive Lyapunov exponent is an indication of chaotic behavior.

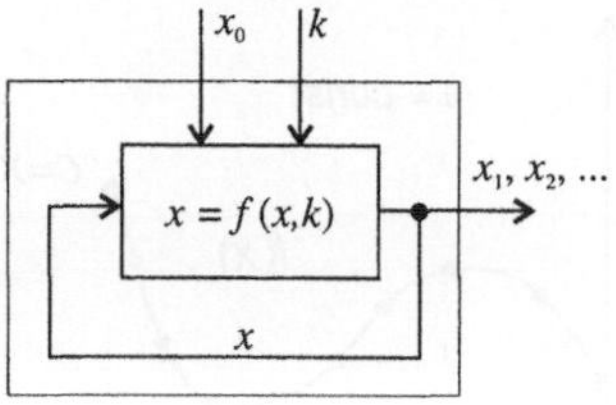

Figure 1. Pseudo-chaos generator based on a iterated function *f* with control parameter *k*.

A simple computer implementation of a chaos generator is a system consisting of a state variable x and an iterated function f (Figure 1). First we seed the system with an $x_0 \in X$ and a control parameter k. Then we perform n iterations $x = f(x)$, during which the sequence of states $x_0, x_1, x_2, \ldots, x_i, \ldots, x_n$, such as $x_i = f^i(x_0)$, is obtained at the output.

This time series appear random and unpredictable to an external observer (e.g. Figure 13). Even if the chaotic map is well defined, prediction becomes difficult when one does not know the seed and can not observe intermediate states. Consequently, chaotic systems can play the same fundamental role as Pseudo-Random Number Generators (PRNG) in conventional cryptography [41, 34]. In particular, major ciphers use PRNG's to produce a sequence of unpredictable cryptographic transformations from a short seed (key).

Section 2 shows that theoretically chaotic systems can be successfully applied to encryption. Unfortunately, we can not directly transfer this result to finite-state approximations. Since the variable x has a finite length and the chaotic map f is approximated, the system is *pseudo-chaotic*, i.e. its properties resemble, but does not coincide with the properties of the original (continuous) chaos. In fact this difference is a critical obstacle, which prevents the developers from using chaos in digital cryptography. Section 4 and 5 is a survey of existing cryptographic systems based on pseudo-chaos.

2 The Concept of Chaotic Cryptography

Chaotic systems can be applied to cryptography in several ways. Firstly, they can act as a component of a conventional cryptographic system, providing the functionality of a PRNG, hash function, permutation block etc. Secondly, chaotic systems can be used as an encryption scheme, for example, symmetric block or stream cipher. This can be achieved with bijective chaotic maps, which ensures inversable, key-dependent and unpredictable transformation of a plaintext message into a ciphertext (Figure 2). Also there is a number of chaos-specific encryption technique based on a single (Figure 3) or multiple trajectories. These methods are described in the next sections of the paper.

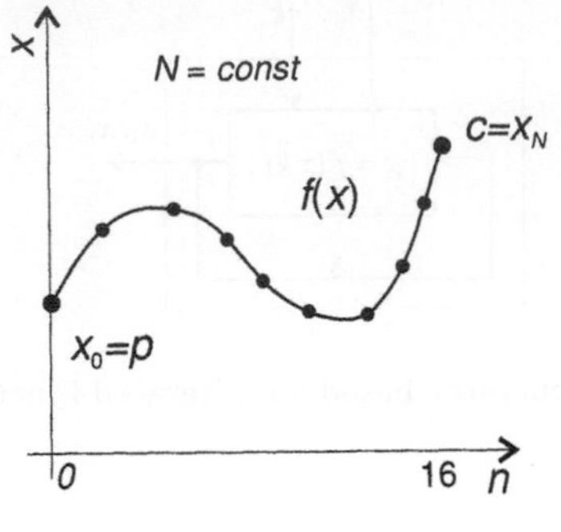

Figure 2. A chaotic block cipher iterates exactly N times. The plaintext p is assigned to the initial state. The ciphertext c is the final state. The key is determined by parameters of the iterated function (not shown on the diagram).

A number generator is a map $G_n : \{0,1\}^n \to \{0,1\}^{p(n)}$, where $p(\cdot)$ is a polynomial satisfying $n+1 \le p(n) \le n^c + c$. An ensemble of generators $\mathcal{G} = \{G_n, n \ge 1\}$ is pseudo-random if: (i) There is a deterministic polynomial-time Turing machine that on input of any n bit strings outputs a string of lenght $p(n)$; (ii) Two probability ensembles $\Pi_1 = G_n(\Pi_0^n)$ and $\Pi_0^{p(n)}$ (the uniform distributions on $\{0,1\}^{p(n)}$) are polynamially indistinguishable for sufficiently large n. Two distributions are polynomial indistinguishable if they assign 'about the same' probability mass to any efficiently recognizable set of strings (formal definitions are given in [15, 27]).

However it's practically impossible to determine whether a generator is pseudo-random using this definition. Numerous statistical tests has been developed to estimate the randomness experimentally (e.g [25, 35]). Some a priori conditions were formulated by Golomb [16]:

The Monobit Test. The number of ones and zeros in a sequence should be approximately the same.

The Runs Test. A run of length k consists of exactly k identical bits and is bounded before and after with a bit of the opposite value. Among all the runs half should be of length $k = 1$, a quarter should be of length $k = 2$, and eighth should be of length $k = 3$ and so on (for each of these lengths there should be equally many runs of zero bits and runs of one bits).

The Autocorrelation Test. The autocorrelation function should be two-valued: when the offset is 0 or is multiple of the period, the value is equal to the period of the generator; otherwise this value is equal to a certain constant integer.

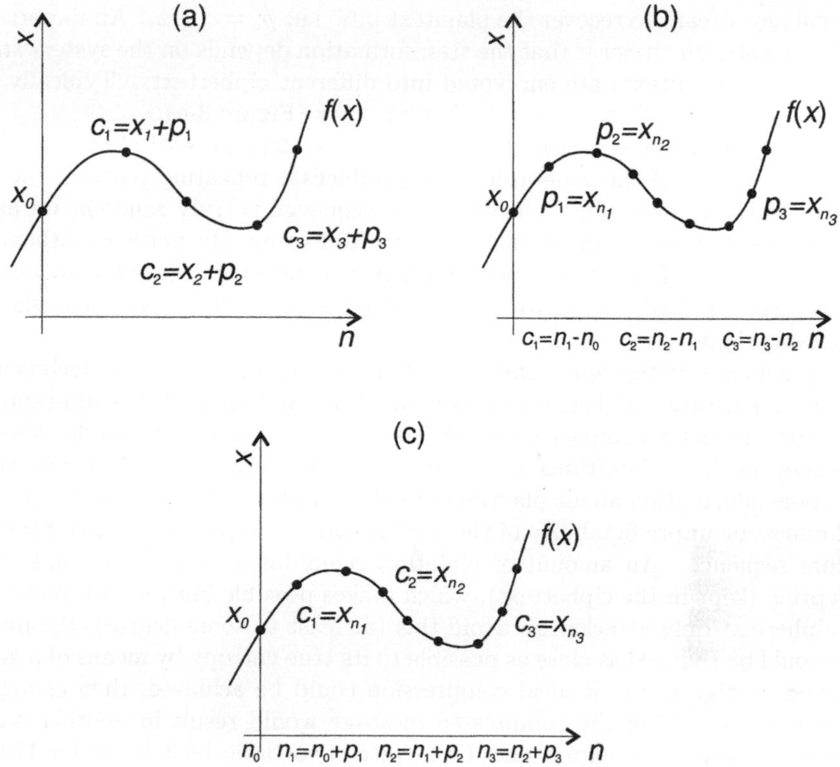

Figure 3. Chaotic stream ciphers based on a single trajectory. (a) The ciphertext c is the plaintext p plus chaotic noise x (e.g. simple XOR cipher using chaos key stream suggested by Bianco); (b) The ciphertext c is a number of iterations n (Baptista); (c) The ciphertext c is the system state after p iterations (Gallagher).

The basic idea of chaotic cryptography is to apply a chaos generator for the purpose of generating pseudo-random numbers. A Pseudo-Chaotic Number Generator (PCNG) is a finite-state approximation of a mathematically chaotic system. The central problem here is under what conditions a pseudo-chaotic sequence is pseudo-random. We can approach to a solution by checking the Golomb properties in PCNG's.

But first we discuss the application of chaotic system to *encrypting transformations.* A simple encryption scheme is a Vernam cipher [34] based on chaotic key-stream (i.e. pseudo-chaotic sequence). A Vernam cipher encrypts a plaintext into a ciphertext 1 bit at a time. A key-stream generator outputs a stream of bits: $x_1, x_2, \ldots, x_n$. This key-stream (sometime called a running key) is XOR'ed with a stream of plaintext bits, $p_1, p_2, \ldots, p_n$ to produce the stream of cipher bits, i.e. $c_i = p_i \oplus x_i$. To decrypt, the ciphertext bits are XOR'ed with an identical key-stream to recover the plaintext bits, i.e. $p_i = c_i \oplus x_i$. An important feature of a stream cipher is that the transformation depends on the system *state* i.e. the same plaintexts are encrypted into different ciphertexts. Typically all the message is encrypted using a single trajectory (Figure 3-a).

The cryptographic strength of a Vernam cipher depends entirely on the randomness property of the generator. If it produces a repeating pattern, the algorithm will have negligible security. If the sequence is truly random, we have the one-time pad cipher and perfect security. Unfortunately, neither mathematical chaos, nor its finite-state approximation can produce a random sequence. By contrast, most of the known chaotic systems generate a highly correlated, predictable signal.

In practical cryptography, one-time pads are not used since it is technically difficult to generate and distribute enormous truly random keys. Pseudo-random generators extract a compact seed into a long sequence. Using pseudo-random sequences, modern algorithms are far from the theoretical perfection and thus keep some information about plaintext in the ciphertext. Roughly speaking, the randomness or unpredictability of the seed is spread through the whole pseudo-random sequence. An amount of plaintext redundancy is still invariant after encryption (kept in the ciphertext), which makes possible both known-plaintext and ciphertext-only attacks. To avoid this (at least to some degree), the plaintext should be reduced as close as possible to its true entropy by means of a good compression algorithm. If ideal compression could be achieved, then changing any number of bits in the compressed message would result in another sensible message when uncompressed. Otherwise, two basic techniques for hiding redundancies in ciphertext should be used (Shannon, [43]):

Confusion ensures that the (statistical) properties of plaintext blocks are not reflected in the corresponding ciphertext blocks. Instead every ciphertext must have a pseudo-random appearance to any observer or standard statistical test.

Diffusion

1. in terms of plaintexts, diffusion demands that (statistically) similar plaintexts do result in completely different ciphertexts even when encrypted with the same key. In particular, this requires that any element of the input block influences every element of the output block in a complex irregular fashion.
2. in terms of a key, diffusion demands that similar keys do result in completely different ciphertexts even when used for encrypting the same block of plaintext. This requires that any element of the key influences every element of the output block in a complex irregular fashion. Additionally, this property must also be valid for the decryption process because otherwise an intruder might recover parts of the

input block from an observed output by a partly correct guess of the key used for encryption.

How are these properties related to chaotic dynamics? An answer can be found in ergodicity theory. Assume that the dynamic system $\mathcal{S} = \langle X, f \rangle$ has a f-invariant measure μ, $\mu(X) < \infty$, i.e.

$$\forall A \in \sigma(X), \qquad \mu(A) = \mu\left(f^{-1}(A)\right),$$

where $\sigma(X)$ is the σ-algebra of measurable subsets of X. Assume the f-invariant measure is equivalent to the Lebesgue with the density function $g(x)$ bounded with some positive constants g_1 and g_2:

$$0 < g_1 < g(x) < g_2,$$

where $\forall A \in \sigma(X)$, $\mu(A) = \int_A g(x)\,dx$. If g_1 is close to g_2 then the measure μ is close to the uniform distribution.

A dynamic system $\mathcal{S} = \langle X, f \rangle$ is ergodic if it has only trivial invariant sets, i.e. if and only if either $\mu(A) = 0$ or $\mu(A \setminus A) = 0$, whenever A is a measurable, invariant under f, subset of the space X (the invariance of A means $f(A) \subset A$). Ergodicity implies that the space X cannot be divided into invariant nontrivial (with respect to the measure μ) disjoint parts (in the case of smaller disjoint parts any brute force attack is restricted to one part of the partition i.e. an intruder will have to search through the whole state space X) [46].

In cryptography we seek for unpredictable systems. A system is computationally unpredictable when there is no efficient algorithm, which predicts the next element of the cryptographic sequence, i.e. all the possible states are equiprobable. If the chaotic map $x_{n+1} = f(x_n)$ is not bijective (e.g. Figure 12), a loss of information occurs during each iteration and one cannot restore the trajectory from x_n to x_0. However, one can use probabilistic techniques to predict the initial conditions. In continuous systems having infinite many states, the uncertainty of x_0 increases to infinity with $n \to \infty$. Consequently, for sufficiently large n the transformation $x_n = f^n(x_0)$ can be considered as a one-way function, which is the essence of conventional PRNG's [15], hash transformations and other cryptographic components.

A dynamic system $\mathcal{S} = \langle X, f \rangle$ is mixing if it satisfies the condition

$$\forall C, P \in \sigma(X), \qquad \lim_{n\to\infty}\left(\frac{\mu\left(f^{-n}(C) \cap P\right)}{\mu(P)}\right) = \frac{\mu(C)}{\mu(X)}.$$

If $\mu(X) = 1$ (the measure μ is probabilistic) we can write

$$\lim_{n\to\infty}\left(\mu\left(f^{-n}(C) \cap P\right)\right) = \mu(C)\,\mu(P).$$

The mixing property implies that the part of P, which after n iterations of f is contained in C is asymptotically proportional to the volume (in the sense of the measure μ) of C in X. Moreover, the iterations of f make each set C statistically independent from P (asymptotically). In other words, the trajectory starting at a fixed point $x_0 \in X$, after n iterations reaches any region with the same probability. Vice versa, for a fixed final state x_n and sufficiently large n, any state x_0 is μ-equiprobable [30].

A simple bit generator G based on a chaotic system $\langle X, f\rangle$ can be constructed in the following way. The state space is divided into two disjoint equiprobable regions X_0 and X_1 i. e. $\mu(X_0) = \mu(X_1) = 1/2$. The regions X_0 and X_1 are assigned to the symbols '0' and '1' respectively. The initial condition $x_0 \in X$ and parameters of f define the seed. During iterations, the state point moves chaotically from one region to another, and the generator produces a symbolic trajectory $\alpha = s_1, s_2, \ldots$, where $s_i \in \{0, 1\}$.

Under the conditions of chaos, ergodicity and mixing (stronger than ergodicity), the system can meet the requirements of several cryptographic applications:

1. Considered as a PRNG $G : X \to \{0,1\}^*$, with the seed the system ensures:

   i. The unique dependence of the sequence $\alpha = G(x_0)$ from the seed $x_0 \in X$. Two different seeds produce two difference sequences [46], i.e

   $$\mu\Big(G^{-1}(\alpha)\Big) = 0.$$

   ii. The equiprobable occurrence of bits in the sequence (The Monobit Test). From Birkhoff-Khinchin Ergodic Theorem the expected number of zeros is equal to the expected number of ones [46]:

   $$\lim_{n\to\infty} \frac{1}{n}\sum_{i=0}^{n-1} \chi_{X_0}\Big(f^i(x)\Big) = \int_{X_0} \chi_{X_0}\, d\mu = \mu(X_0), \tag{2.1}$$

   $$\lim_{n\to\infty} \frac{1}{n}\sum_{i=0}^{n-1} \chi_{X_1}\Big(f^i(x)\Big) = \int_{X_1} \chi_{X_1}\, d\mu = \mu(X_1), \tag{2.2}$$

   where χ_{X_0}, χ_{X_1} are indictor functions of the subsets X_0 and X_1 respectively. Since $\mu(X_0) = \mu(X_1) = 1/2$, the average number of zeros and ones tends to $n/2$.

   ii. The asymptotic statistical independence of bits (The Autocorrelation Test). For all seeds $x_0 \in X$ and all $n = 1, 2, \ldots$ the bits $s_{(n-1)k}$ and s_{nk} considered as random variables) of the symbolic trajectory $\alpha = G(x_0)$ are asymptotically independent as k increases [46], i.e.

   $$\lim_{k\to\infty} \mu\Big(f^{-nk-k}(X_0) \cap f^{-nk}(X_1)\Big) = \mu\Big(f^{-nk}(X_0)\Big) \cdot \mu\Big(f^{-nk}(X_1)\Big)$$

   and

   $$\lim_{k\to\infty} \mu\Big(f^{-nk-k}(X_1) \cap f^{-nk}(X_0)\Big) = \mu\Big(f^{-nk}(X_1)\Big) \cdot \mu\Big(f^{-nk}(X_0)\Big).$$

2. Considered as an encryption scheme, the chaotic system ensures:

   i. Confusion (statistical properties of the plaintext are not reflected in the ciphertext, similiar to the PRNG application); and

   ii. Diffusion. In terms of the plaintext, diffusion is achieved when the plaintext is introduced in the initial condition, like in any block cipher. In terms of the key, diffusion takes place when the key is assigned to a control parameter or is mixed with the plaintext (e.g. using XOR). The sensitivity to the initial conditions is closely related to diffusion. Both in chaos and cryptography we are dealing with systems in which "*a small variation of any variable changes the outputs considerably*" [43].

The Lyaponov exponent (1.1) is a quantitative measure of the sensitivity to the initial conditions. In chaotic encryption, λ is the measure of cryptographic efficiency. Larger λ indicates that fewer chaotic transformation will be required to reach a certain level of the ciphertext entropy. Numeric values of λ are given for a few encryption systems described in the next sections of the paper. Also, the deterministic dependence of the cryptographic sequence from the seed provides key-dependent encryption and decryption.

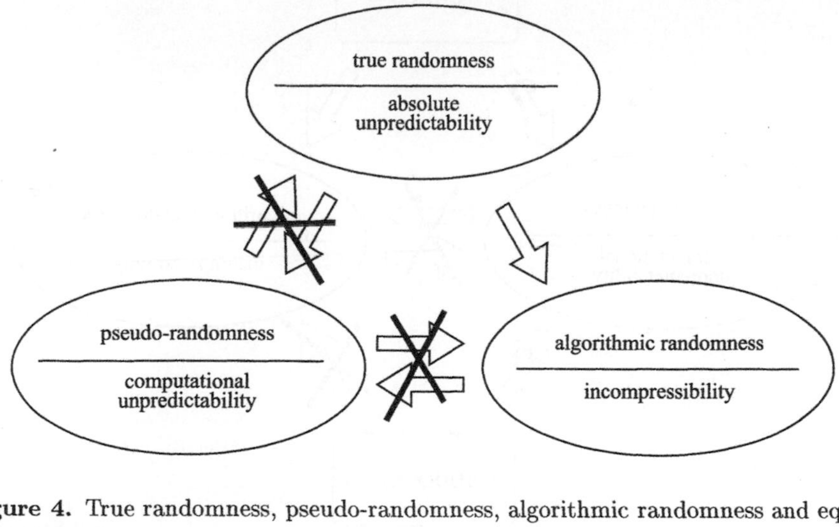

Figure 4. True randomness, pseudo-randomness, algorithmic randomness and equivalent concepts.

Figure 4 describes the relationship between fundamental properties of a cryptographic object. The perfect security is achieved at the theoretical layer, where the object is absolutely unpredictable or truly random. Randomness is associated with the uniform distribution on the entire state space of the object. A microscopic parameter of a high-dimension chaotic system, like ideal gas, is an infinite source of information and generates a truly random signal. There are real world applications using natural chaos in cryptography, for example, Intel's hardware RNG captures randomness from the thermal noise of the environment. However, such random sequences are used only in key generation, not in encryption, because there are not reproducible.

At the practical layer we deal with 'pseudo-' concepts. Pseudo-random object cannot be efficiently distinguished from the uniform noise (there is a mathematical definition of a pseudo-random object in terms of complexity theory [15]), there are produced by a compact generator.

The object is algorithmic random if its length equals the length of the shortest program (algorithm) generating the object. Algorithm random objects are incompressible, i.e. have no redundancy and do not display patterns.

Clearly, a truly random is object is algorithmically random, but not inversely: an algorithmically random object is not truly random and it can be predicted by a probabilistic machine. Pseudo-random object is not algorithmic random because it is efficiently compressed by the generator; however it is computationally incompressible for an external observer.

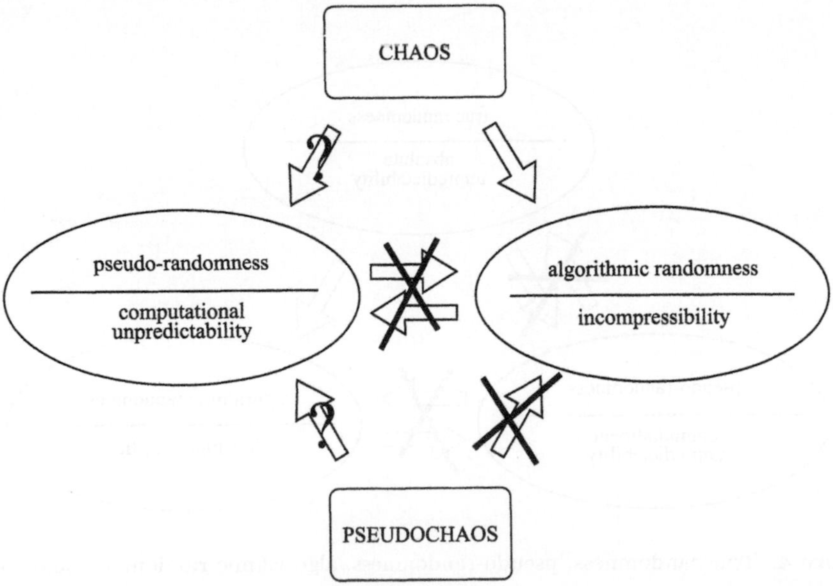

Figure 5. The relationship between chaos and cryptographic concepts.

Figure 5 illustrates the relationship between chaos and cryptographic concepts. Brudno's and White's theorems (cited in [4, 27]) show that almost all symbolic trajectories of a chaotic system are algorithmic random, whereas pseudo-chaos, a finite-state approximation of chaos, is not.

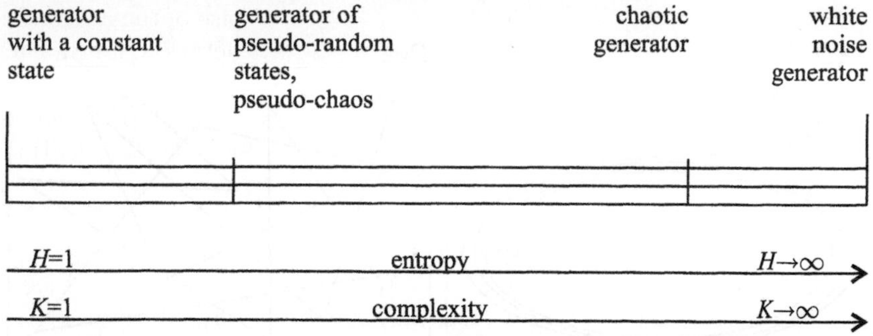

Figure 6. The randomness of cryptographic and chaotic systems.

Randomness can be 'measured' using well known properties: algorithmic complexity (the length of the shortest program producing the object) and Shannon entropy (the degree of our uncertainty about the object). The two concepts a related as 'cause–effect': the more the complexity of the object, the more is its entropy. Quantitatively, Shannon entropy is in direct proportion to the algorithmic complexity (in ergodic systems, where statistical properties of a single sequence coincides with that of all sequences, emitted by a generator) [4].

'Fully predictable' object has a well-know state, its complexity is 1. Fully unpredictable object (white noise) is an infinite source of information and has and the infinite complexity (Figure 6). Cryptographic systems are somewhere in-between: there are complex enough to be unpredictable by an external observer, but not too much to be reproducible.

Chaotic systems are at the end of this scale:

- High-dimension chaos has infinitely many states and infinitely many independent variables. It can have the highest level of the entropy of the white noise generator, as we say, but generally its entropy is considerably lower because of the correlations.

- Low-dimension chaos has infinitely many states and a small number of independent variables. The uncertainty of a sequence, generated by a low-dimension chaotic system with a known iterated function, lies in the initial conditions. The computational unpredictability depends entirely on the iterated function. Certain iterated functions with unknown parameters can provide the unpredictability and the uniform distribution of the sequence.

There are fundamental differences between chaos theory and cryptography.

Firstly, chaotic systems are defined on infinite state spaces, whereas cryptography deals with a finite number of binary string. Pseudo-chaos, a finite-state approximation of true chaos, in most case differs essentially from the original system and is much more predictable.

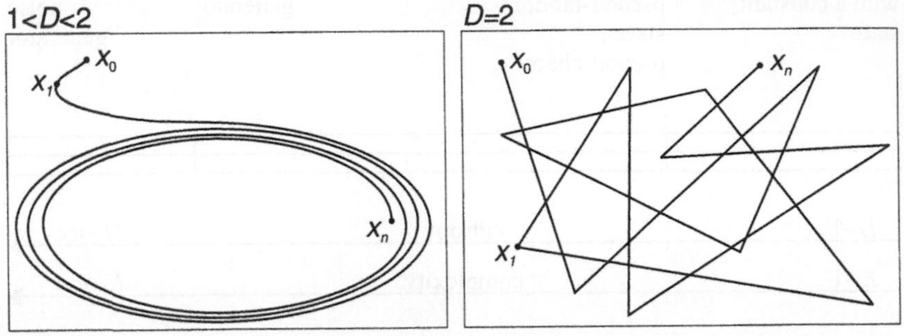

Figure 7. Phase space diagrams for chaotic and cryptographic systems. Classical chaotic systems have a distinguishable attractor with a fractional dimension (left). Cryptographic systems attempt to hide any order of the phase space, maximizing the dimension to the number of variables (right).

Particularly, chaos theory studies the phase-space portrait of a continuous system, strange attractors and repellors. An attractor has a complex and bounded shape with a fractional dimension, so the system behavior is limited by a certain configuration in the state space. Cryptography attempts to make the phase-space picture most unpredictable, avoiding recognizable attractors (i.e patterns). This can be achieved, when all possible configurations are equiprobable and the dimension equals the number of independent variables (Figure 7).

Another difference is that the theory of dynamic systems aims to understand the asymptotic behavior of iterative processes ($n \to \infty$). In cryptography one focuses on the properties of:

1. the first iteration ($n = 1$), if a chaotic system describes a single cryptographic transformation of a cipher, or
2. a small number of first iterations (often $n = 16$), if it describes a round transformation in multi-round encryption scheme, or
3. multitude of iterations ($n > 2^{10}$), if it describes a stream cipher or a PRNG.

A pseudo-chaotic system can produce cycles with different lengths, sometimes very short (Figure 8-a). This leads to recognizable patterns, which is not expectable in cryptography. A system with a single orbit (Figure 8-b) is the best choice for a long run PRNG; it provides the maximum length of an unpredictable

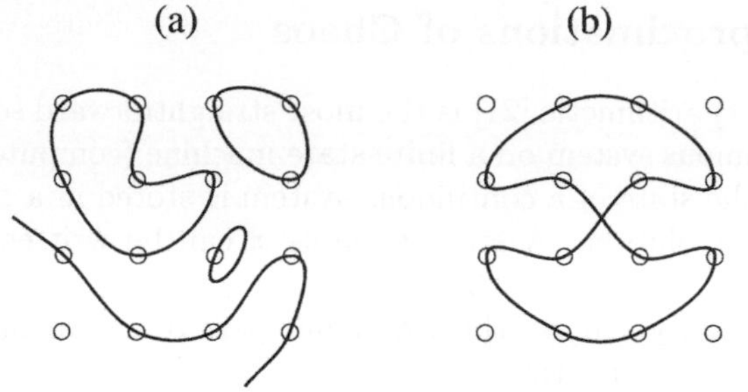

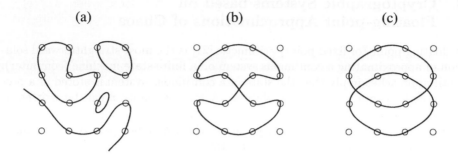

Figure 8. Abstract orbits of a pseudo-chaotic system. (a) Dangerously short and unpredictable orbits (unsuitable for cryptography); (b) A single orbit (suitable for a long run PRNG or a cipher); (c) Multiple orbits with the same length (suitable for a block cipher).

string. However, systems with multiple orbits of the same length (Figure 8-c) can be used in a block cipher or other cryptographic components, where the number of rounds is small.

In this paper we consider two classes of pseudo-chaos. The first class represents chaotic systems approximated with floating-point numbers; the second one is systems defined on a state space of binary strings. Though both computer implementations are eventually based on binary arithmetic, the difference is that in the floating-point approach we measure the floating-point distance between states and evaluate the iterated function with floating-point calculations, whereas in binary chaos we have the Hamming distance and binary chaotic functions.

3 Continuous Cryptographic Systems

Numerous investigators have studied the application of chaos to analog coding and encrypting. Caroll *et al.* [7], Kocarev *et al.* [26], Dachselt [10], Chu *et al.* [9], Sobhy *et al.* [44] and Wu *et al.* [52] describe encryption schemes based on synchronized chaotic circuits. Notably, the book *Chaotic Electronics in Telecommunications* by Kennedy *et al.* [24] is the first fundamental description of industry-ready chaotic techniques. These analog schemes belong more to the field of radio communication and hardware encryption, which are not discussed in this paper.

4 Cryptographic Systems based on Floating-point Approximations of Chaos

A floating-point (or fixed point) arithmetic [21] is the most straightforward solution of approximating a continuous system on a finite state machine (computer). Both approaches imply that the state of a continuous system is stored in a program variable under a finite resolution. A state variable x can be written as a binary fraction $b_m b_{m-1} \dots b_1 . a_1 a_2 \dots a_s$, where a_i, b_j are bits, $b_m b_{m-1} \dots b_1$ denotes the integer part and $a_1 a_2 \dots a_s$ is the fractional part of x. In finite resolution, instead of $x_{n+1} = f(x)$, we write

$$x_{n+1} = round_k (f(x_n)),$$

where $k \leq s$ and $round_k(x)$ is a rounding function defined as

$$round_k (x) = b_m b_{m-1} \dots b_1 . a_1 a_2 \dots a_{k-1} (a_k + a_{k+1}).$$

Floating-point encryption systems require a mapping from the plaintext alphabet (e.g. 8 bit symbols) to the state space (64 bit floating-point numbers) and/or from the state space to the ciphertext alphabet. The set of m disjoint regions $\beta = \{X_1, X_2, ..., X_m\}$ which covers the continuous state space X is called a partition:

$$X = \{X_1, X_2, ..., X_m\} : \bigcup_{i=1}^{i=m} X_i = X \text{ and } X_i \cap X_j = \emptyset \text{ for all } i \neq j.$$

A unique symbol $s = \sigma(X_i)$ is assigned to every region $X_i \in X$. The process of partitioning the state space, assigning symbols to every region from the partition, and the resulting macroscopic dynamics are called symbolic dynamics. In cryptographic applications, we can apply the same technique, assigning partitions the plaintext or ciphertext symbols. However, it is *not* obligatory to occupy all the partitions and assign unique symbols. On the contrary, we can change statistical properties of the resulting symbolic trajectory by assigning symbols in a complex manner.

The central problem is that $round_k(x)$ is applied in each iteration and the rounding off accumulation causes the original and the approximated processes to diverge very fast. As it was discovered by Lorenz[2], " *... a small error in the former will produce an enormous error in the future. Prediction becomes impossible ...*". Floating-point arithmetic is a bad approximation because the approximated model does not converge asymptotically with the continuous model, and, furthermore, exhibits nonchaotic properties, e.g. cycles and patterns (for example, Figure 8-a).

For cryptographic applications, the rounding off function exposes another danger. Rounding or truncating of the state can lead the process out of the

[2]Lorenz, Edward, is an American meteorologist, who discovered a stable chaotic attractor in 1960s.

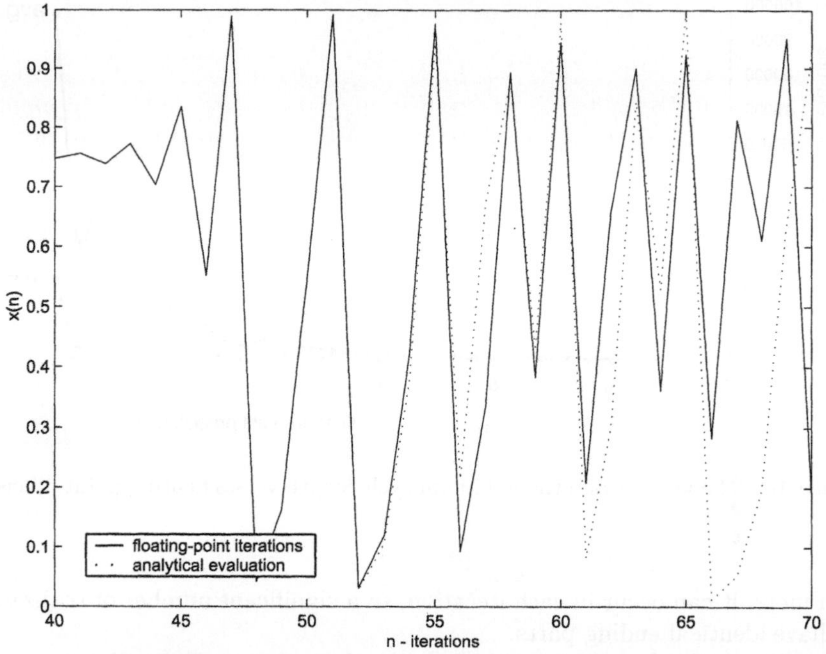

Figure 9. Trajectories of a continuous chaotic system (the logistic map) and its floating-point approximation. The rouning off error is amplified in each iteration and the trajectories diverge exponentially. The continuous system is evaluated using an analytical solution.

chaotic attractor. After that, the system state typically directs to a certain constant value or infinity. So we have to exclude some forbidden initial conditions and parameters, which cause a constant or pattern behavior after a small number of iterations. Figure 9 shows how suddenly the original and approximated trajectories diverge, and Figure 10 is a plot of the average cycle length verses floating-point precision.

Another problem is the sensitivity to floating-point processor implementations. Diversified mathematical algorithms or internal precisions in intermediate calculation can lead to the situation when the same code of an encryption software will generate different cryptographic sequences.

In section 2 we have mentioned that a PCNG with two different seeds produces two different sequence with probability 1. This is true for chaotic systems with an infinite state space, where the probability $\Pr\left(f(x_n), f(x'_n)\right) \to 0$ with $x_n \neq x'_n$ (despite of the nonlinearity of the map f). In finite-state approximation, the probability of mapping two points into one is much higher (e.g. Figure 12).

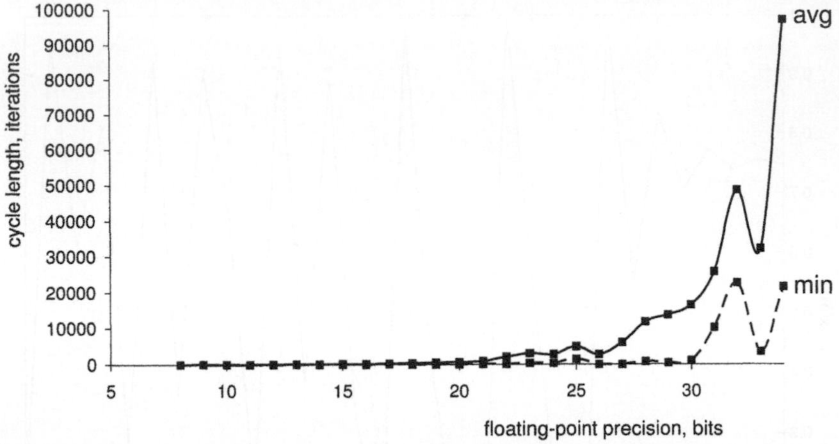

Figure 10. The average and the minimum cycle length verses floating-point precision.

Even more, it can occur in each iteration, so a significant number of trajectories will have identical ending parts.

In spite of the rounding problems, a number of investigators have explored applications of continuous chaotic to digital cryptography. Following is a publication survey on encryption schemes based on the floating-point approximation of chaos.

In 1983 Erber *et al.* [11] suggested using a Chebyshev mixing polynomial to simulate a random process on digital computers. The polynomial is given by

$$x_{n+1} = x_n^2 - 2, \qquad \text{where } x_n \in (-2, 2)\,.$$

The sequence $x_0, x_1, x_2, \ldots$ is chaotic, however, some particular values for x_n must be avoided ($x_n \neq -1, 0, 1$). With a non-integral seed x_0, it is theoretically expected that these values will never be encountered. A system with 40-bit precision provided the average cycle of 105 steps.

The output y_n is obtained after an additional nonlinear transformation

$$y_n = \frac{4}{\pi} \arccos\left(\frac{x_n}{2}\right) - 2.$$

This additional transformation attempts to improve the statistical properties of the sequence, which has a particular probability distribution (the same as for the logistic map given in Figure 16). Cryptographic applications require pseudo-random sequences (i.e. sequences, which are computationally undistinguishable from the uniform distribution). By applying a transformation with the same shape as the Cumulative Distribution Function (CDF) of the input sequence, one can obtain a uniform-like CDF of the output (Figure 11). However the

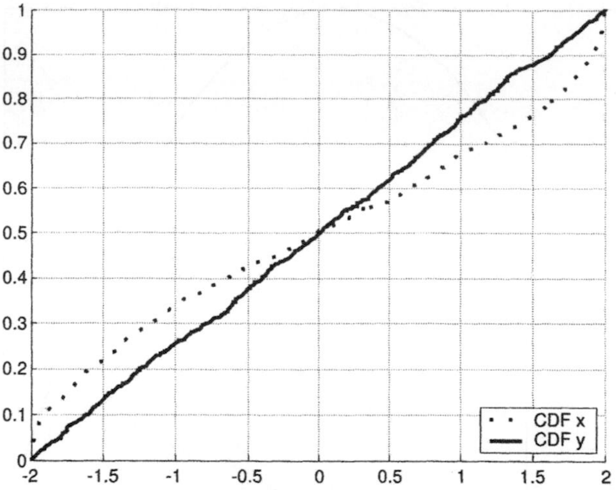

Figure 11. The empirical Cumulative Distribution Function of a sequence generated with the Chebyshev system. The straight line looks like the uniform distribution.

transformation does not affect the *order* of elements, which is also important. For instance, the run test of randomness consists in the following: among all the runs half should be of length 1, a quarter should be of length 2, an eighth should be of length 3 and so; for each of these lengths there should be equally many runs of zero bits and runs of one bits. The run test for Chebyshev map and similar systems fails.

So far, several investigators (Hosack [22], for example) have found that the Chebyshev transformation possesses undesirable qualities which make it unsuitable for pseudo-random number generation.

A similar transformation became the most famous chaotic map. Earlier, in 1976, Mitchell Feigenbaum studied a complex behavior of the so-called logistic map (Figure 12)

$$x_{n+1} = 4rx_n\left(1 - x_n\right), \tag{4.1}$$

where $x \in (0,1)$ and $r \in (0,1)$.

With certain values of parameter r, the generator delivers a sequence, which *appears* pseudo-random (Figure 13). The Freigenbaum cascade (Figure 14) shows the values of x_n on the attractor for each value of the parameter r. As r increases, the number of points in the attractor increases from 1 to 2, 4, 8 and infinity. In this area ($r \to 1$) it was considered difficult to estimate the final state of the system (without performing n iterations) given initial conditions x_0, or vice-versa - to recover x_0 (which can be a key or a plaintext) from x_n. This complexity was regarded as a fundamental advantage in using continuous chaos for cryptography.

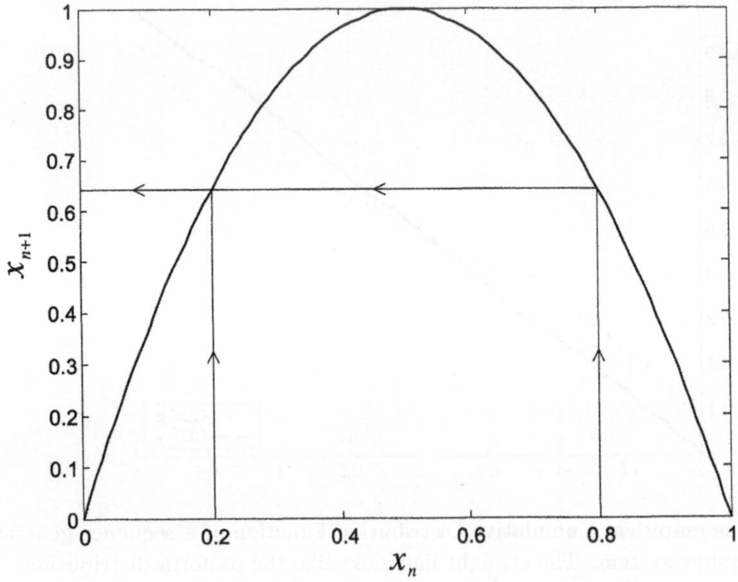

Figure 12. The logistic map for $r = 0.99$.

However, recent developments in chaotic dynamics allows one to evaluate exactly x_n from x_0 [30].

For any long sequence of N numbers generated from the logistic map we can calculate the Lyapunov exponent given by

$$\lambda(x_0) = \frac{1}{N}\sum_{n=1}^{N} \log|r(1-2x_n)|.$$

The numerical estimation for $r = 0.9$ and $N = 4000$ is $\lambda(0.5) \approx 0.7095$.

Matthews [33] generalizes the logistic map with cryptographic constraints and develops a new map, which exhibits chaotic behavior for parameter values within an extended range (Figure 15). The Matthews map is given by

$$x_{n+1} = (r+1)\left(\frac{1}{r}+1\right)^r \cdot x_n(1-x_n)^r, r \in (1,4).$$

He suggests using this map for generating a sequence of pseudo-random numbers which can be used as a one time pad for encrypting messages.

Bianco *et al.* [3] uses the logistic map (4.1) to generate a sequence of floating point numbers, which is then converted into a binary sequence. The binary

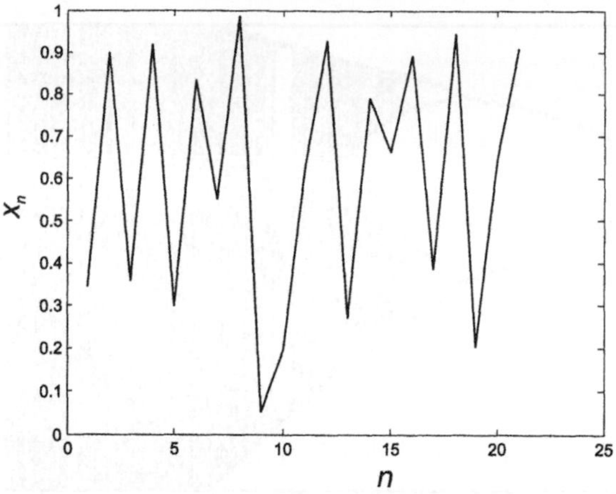

Figure 13. A chaotic sequence generated with the logistic map for $x_0 = 0.34$ and $r = 0.99$.

sequence is XOR-ed with the plain-text, like in the one-time pad cipher. The parameter of the logistic map together with the initial condition form part of the ciphering key. The conversion from floating point numbers to binary values is done by choosing two disjoint interval ranges representing 0 and 1. The ranges are selected in such a way, that the probabilities of occurrence of 0 and 1 are equal (Figure 16, for example). Note, the equiprobable mapping does not ensure the uniform distribution. Though the numbers of zeros and ones are equal, the order is not random.

It has been pointed out by Wheeler [48] and Jackson [23] that computer implementations of chaotic systems yields to surprisingly different behavior, i.e. it produces very short cycles and trivial patterns (a numeric example in this paper is given in Figure 10).

The one-dimension tent map is defined as

$$x_{n+1} = \begin{cases} x_n/a, & 0 \leq x_n \leq a \\ \frac{1-x_n}{1-a}, & a < x_n \leq 1 \end{cases}, \tag{4.2}$$

where the parameter a determines the top of the tent (Figure 17).

The Lyapunov exponent depends on the parameter a and is given by $\lambda(a) = -a\ln(a) - (1-a)\ln(1-a)$ for almost all $x_0 \in (0,1)$ [31].
Numerically, $\max\limits_{a\in(0,1)} \lambda(a) \approx 0.693$ at $a = 0.5$.

Habutsu *et al.* [19] suggested a chaos-based block encryption system (Figure 2) in which the *inverse* of the one-dimensional tent map (4.2) is applied $N = 75$ times to an initial condition representing the plaintext (Figure 2). A combination

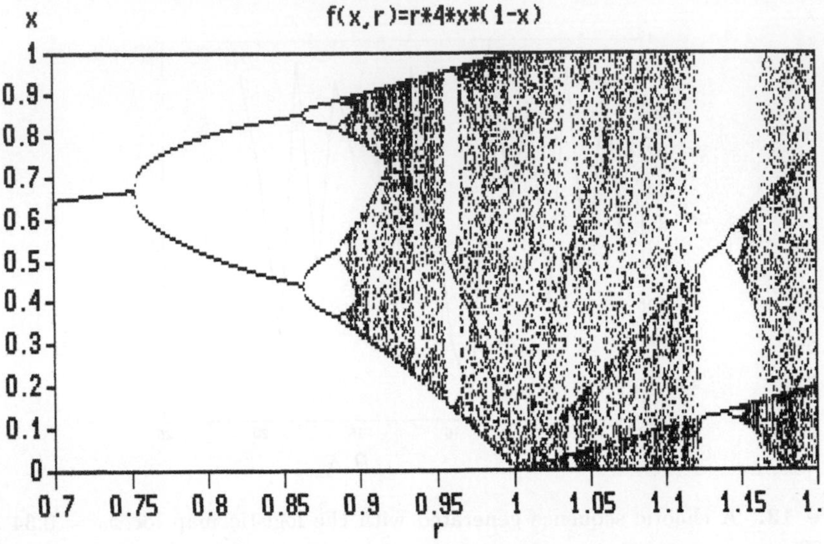

Figure 14. Bifurcation of the logistic map. The most 'unpredictable' behavior occurs when $r = 1$.

of a floating-point number $a \in [0.4, 0.6]$ and a binary sequence $r = \{r_n\}_{n\in[1,N]}$ is the secret key. The map can be written as

$$x_{n+1} = \begin{cases} ax_n, & r_n = 0 \\ (a-1)\,x_n + 1, & r_n = 1 \end{cases}, \qquad n \in [1, N]. \tag{4.3}$$

For N inverse iterations there are 2^N possible ciphertexts which encode the same plaintext. Biham [2] pointed out that the cryptosystem could be easily broken using a chosen ciphertext type of attack; the complexity of a known plain-text type of attack is 2^{38}.

Protopopescu [37] proposes an encryption scheme based on multiple iterated functions: m different chaotic maps are initialized using a secret key. If the maps depend on parameters, these too are determined by the key. The maps are iterated using floating point arithmetic and m bytes are extracted from their floating point representations, one byte from each map. These m numbers are then combined using an XOR operation. The process is repeated to create a one time pad which is finally XOR-ed with the plain-text.

Gallagher *et el.* [14] developed a chaotic stream cipher, based on the transformation

$$f(x) = \left(a + \frac{1}{x}\right)^{\frac{x}{a}} \qquad \text{where } x \in (0, 10) \text{ and } a \in [0.29, 0.40].$$

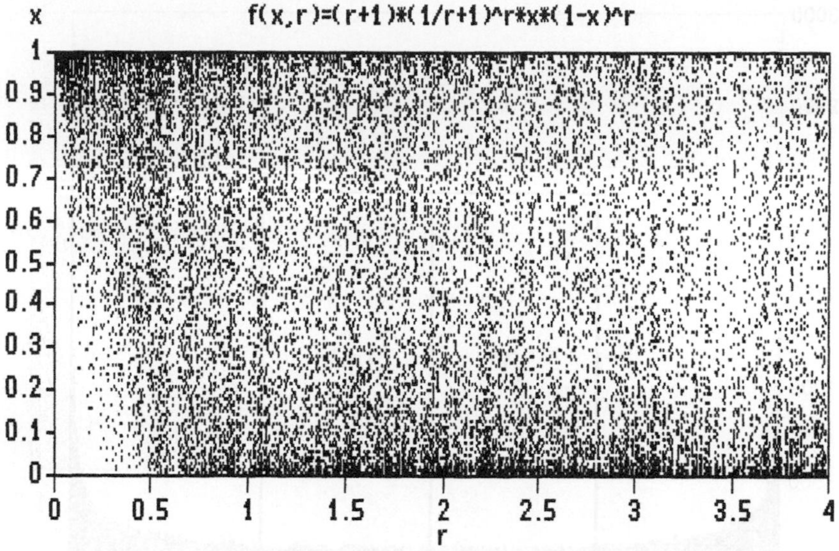

Figure 15. Attractor points corresponding to different values of the parameter r in the Matthews map.

Both the initial condition x_0 and the parameter a represent the key. After $n_0 = 200$ begining iterations, the system encrypts the plaintext byte p_1 into the ciphertext float $c_1 = f^{n_0+n_1}(x_0)$, i.e. the chaotic map is applied $p_1 \in [0, 255]$ times. Subsequent plaintexts are encrypted using the same trajectory (Figure 3-c):

$$c_i = f^{\sum_{i=0}^{p_i} n_i}(x_0).$$

Visibly, disadvantages of such an encryption scheme are (i) the data expansion (the floating-point representation of c_i is considerably larger that the source byte p_i) and (ii) unstable cycles, incident to floating-point chaos generators.

Baptista [1] and Wong [51] propose a few encryption methods in which the ciphertext is *a number of iterations*. The state space X of a chaotic system is partitioned into m disjoint regions $\{X_1, X_2, ..., X_m\}$ (covering the entire X or, possibly, part of it). A unique plaintext symbol p is assigned to each region. Seeded with an initial condition (x_0, control parameters), the system performs n_0 iterations. Then all the plaintexts are encrypted into a sequence of integers

$$c_i = n_i - n_{i-1}, \quad i = 1, 2, \ldots,$$

where $p_i = \sigma(X_i)$, $f^{n_i} \in X_i$. Figure 3-b shows the concept of iteration counting ciphers. This encryption scheme can be extended with a pseudo-random n_{min},

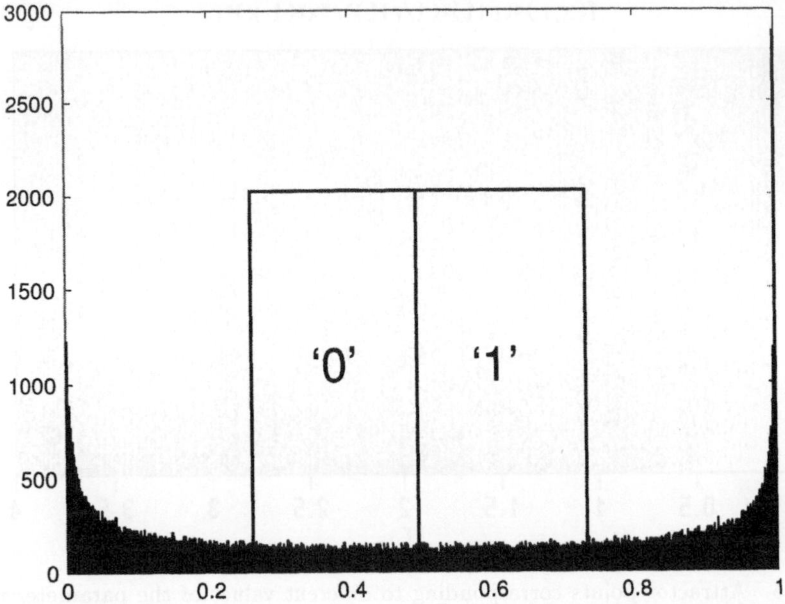

Figure 16. A Probability Distribution Function of a chaotic sequence. The output bins '0' and '1' are equiprobable, but correlated.

generated for each plaintext symbol. The integer n_{min} defines the minimum number of iteration the system should perform before encrypting the next symbol. Clearly, the disadvantages are: (i) a probability of overflowing the counter $c_i = n_i - n_{i-1}$ (i.e. the length of a trajectory segment can be larger then the maximum number fitting the fixed-length ciphertext c_i); (ii) the ciphertext being larger than the plaintext; (iii) lots of redundant cycles when long orbits occur.

Ho [20] continued the research of Baptista and Wong, and went on with a software implementation of the encryption techniques, performance evaluation and cryptanalysis. Ho suggests using 4 plaintext bit blocks to speed up the cryptographic algorithm. However, this increases ciphertext redundancy — 4 bits of plaintext is encrypted into 10 bit ciphertext block. He discusses various chaos-specific cryptography issues — dynamic mapping from plaintext to floating-point state, trajectory distribution and algorithm complexity.

Kotulski [28, 29] introduces a secret key into the initial conditions and parameters of the floating-point approximation of chaotic system. Part of the initial conditions is a plaintext; the rest contains a key. The ciphertext is the system state after a fixed number of iterations (Figure 2). In his Discrete Chaotic Cryptography method Kotulski proposes a two dimensional map matching the

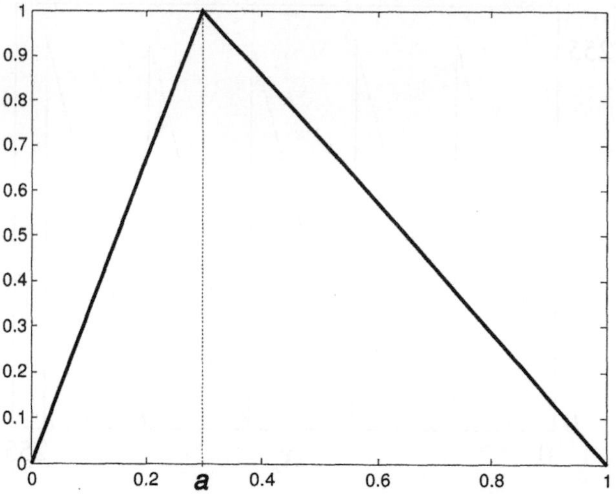

Figure 17. The tent map.

reflection law of a geometric square and defines conditions under which the system is chaotic and mixing.

To increase unpredictability (i.e. the number of states, nonlinearity, complexity) high-order multi-dimension chaotic system can be used. For example, Paar [36] suggests a second order differential equation describing a robot model

$$m\frac{d^2x}{dt^2} - \beta_2\left(\frac{dx}{dt}\right)^2 sign\left(\frac{dx}{dt}\right) - \gamma_2 \mathrm{sign}\left(\frac{dx}{dt}\right) - \delta_{21}x - \delta_{23}x^2 =$$

$$= L\frac{\omega_0^2}{2\pi}\cos(\omega_0 t) - \zeta_{21}e^{\lambda_{21}t} - \zeta_{22}e^{\lambda_{22}t},$$

which correspond to a hard spring with coefficients δ_{21} and δ_{23} with two types of friction. The right side of the equation represents a periodic time-dependent force with the amplitude $L\left(\omega_0^2/2\pi\right)$ and a feedback force given by the corresponding parameters.

To date, no such systems have been implemented as a working encryption algorithm (at least no algorithm is known to the author). This is principally due to the relatively complex numerical integration schemes that are required and the non-uniform distribution of state variables. Cryptography needs more efficient solutions.

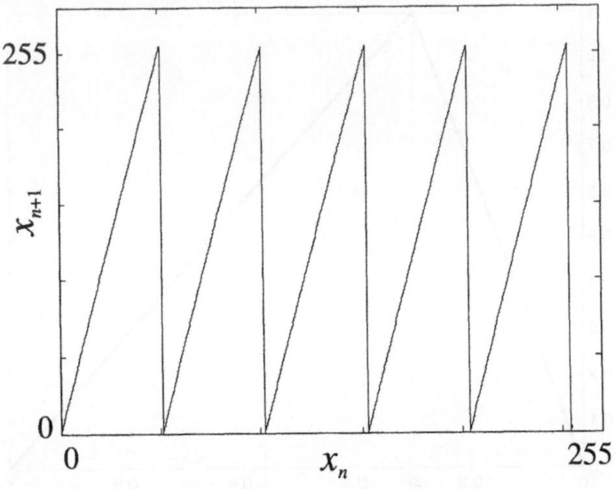

Figure 18. The sawteeth map for $p = 1279$ and $q = 255$.

4.1 Unpredictable and Solvable Chaotic Systems

Consider the sawteeth map

$$x_{n+1} = rx_n \bmod q,$$

where $x_0 \in [0, q]$, $r = p/q > 1$ and p is a co-prime (relative prime) of q. The map is chaotic for all r and has a Lyapunov exponent $\lambda = \log r > 0$ [27]. The chaotic property of the sawteeth map is widely used in conventional cryptography, for example, the popular Linear Congruental Generator (LCG) is given by $x_{n+1} = (Ax_n + C) \bmod M$, where $x_n \in \{0, 1, \ldots, M\}$ and A, C, M are integers fixed by the designer (Knuth, [25]). Using our vocabulary, the LCG is a typical pseudo-chaotic binary system (described in the next section).

It's easy to show that the sequence $y_1, y_2, \ldots, y_n, \ldots$ is chaotic if $y_n = H(x_n)$ and $H : R \to [0, 1]$ is a periodic function with period 1 [27], e.g. $H(x) = \sin^2(\pi x)$.

For large p and q, the sequence $\{y_i\}_1^\infty$ is *one-step unpredictable*: for any element y_n in the sequence $\{y_i\}_1^\infty$ one can only guess among q equally distributed values the next element $y + n + 1$ and among p equally distributed values the previous element y_{n-1} [27].

A solvable system has an analytical solution of each trajectory, i.e. for any discrete moment of time n we can calculate the state x_n from the initial condition x_0 (without performing n iterations). For example, for this logistic system there is an analytical solution

$$x_n = \sin^2\left(2^n \arcsin \sqrt{x_0}\right). \tag{4.4}$$

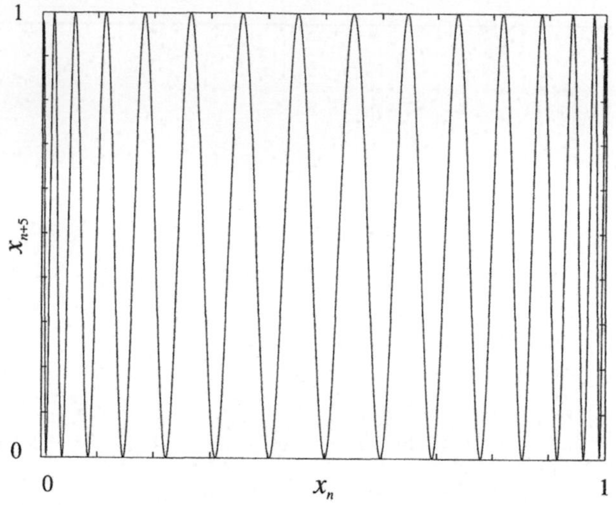

Figure 19. The state of the logistic system after $n = 5$ iterations versus the initial condition.

For $n = 1$ this equation is equivalent to the logistic map (4.1). Figure 19 is a plot of $x_5 = f^5(x_0)$. If n increases, the number of peaks will grow illustrating the sensitivity to the initial condition.

A solution for the tent map (4.2) is given by

$$x_n = \frac{1}{\pi} \arccos\left(\cos 2^n \pi x_0\right). \tag{4.5}$$

The solutions (4.4) and (4.5) lead directly to a number of new solvable chaotic systems [30]. Clearly, the exact solution of a dynamic system is important in simulation sciences since it eliminates the rounding accumulation problem and speeds up computations. A special class of solvable systems is very interesting for cryptographic applications. These are one-step unpredictable solvable systems, described by Gonzalez *et al.* [17].

All known solutions for a chaotic systems can be written in the general form

$$x_n = \Psi\left(\theta T \kappa^n\right), \tag{4.6}$$

where $\Psi(t)$ is a periodic function with period T, κ is an integer and θ is a real parameter defining the initial condition:

$$x_0 = \Psi\left(\theta T\right).$$

The system is chaotic if the Lyapunov exponent $\lambda = \ln \kappa > 0$.

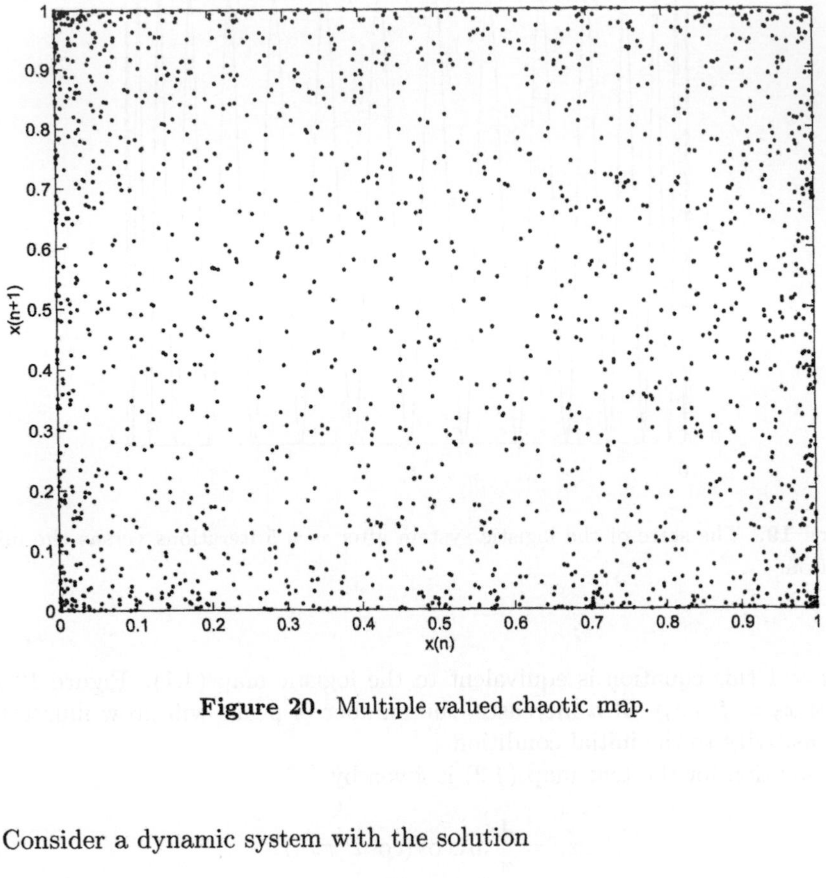

Figure 20. Multiple valued chaotic map.

Consider a dynamic system with the solution

$$x_n = \sin^2\left(\pi\theta\kappa^n\right). \tag{4.7}$$

The one-step map $x_n \to x_{n+1}$ for this system is given by

$$x_{n+1} = \sin^2\left(\kappa \arcsin\sqrt{x_n}\right). \tag{4.8}$$

Unlike the logistic and the tent equations, the map (4.8) can be *multi-valued*, in particular, when κ is an irrational number. Figure 20 shows an undefined set of points obtained with (4.7) for $\kappa = \pi^{1/3}$.

Kotulski *et al.* [30] studied chaos generators based on the unpredictable map (4.8) and found them useful for producing short chaotic sequences (e.g. in block ciphers). Long-run generators (e.g. in steam ciphers) need additional research because floating-point implementations are very sensitive to the choice of seeds and control parameters, i.e. not all seeds produce a sufficiently long sequence.

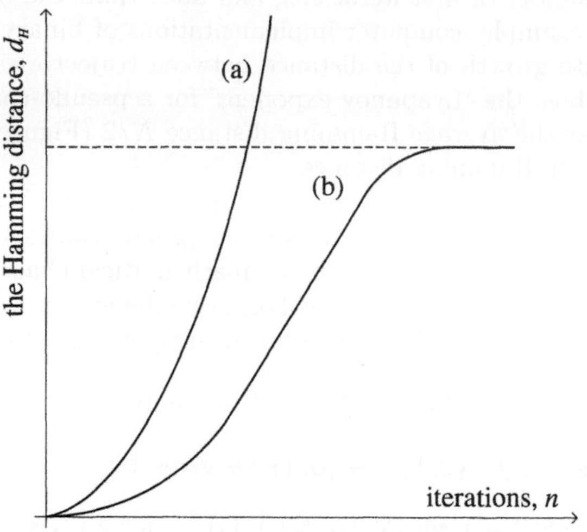

Figure 21. The Lyaponov exponent of a chaotic system (a) and the analogous property of a finite-state pseudo-chaotic system (b).

5 Binary Chaotic Systems

In 1998 Waelbroeck and Zertuche [47] propose a theory of deterministic chaos for binary systems. Binary systems, like a cellular automata and neural networks, are the best known class of chaotic discrete systems.

The state space $\Omega = \{\omega_j\}$ is the Cartesian product of infinite copies of N-bit binary sequences $s = \{0,1\}^N$.

By introducing a topological base $\mathcal{N}_n(\omega)$, similar to the Hamming distance, the space Ω becomes homeomorphic to the space of 2^N symbols in symbolic dynamics. The two basic properties of chaos — sensitive dependence on initial conditions and topological transitivity — can be naturally generalized from subsets of R^N to Ω. The Hamming distance $d_H(\omega_1, \omega_2)$ is a measure of difference between two states, specifically, the number of bits that differ.

This definition of Lyapunov exponent (1.1) applies to continuous-state systems. In binary chaos, the value of ε is limited downwards by one bit, but the distance (i.e. divergence) between states can grow infinitely. A common definition of the Lyapunov exponent for binary systems is given by

$$\lambda(\omega_0) = \lim_{n\to\infty} \frac{1}{n} \log d_H\left(f^n(\omega_0), f^n(\omega_0')\right)$$

where $\omega_0' \in \Omega$ such as $d_H(\omega_0, \omega_0') = 1$.

However, the space Ω is infinite, while cryptographic applications are based on finite-state devices. So binary pseudo-chaos can exhibit chaotic properties during a finite number of first iterations, and after that, the behavior becomes nonchaotic. For example, computer implementations of binary systems can not provide an infinite growth of the distance between trajectories, starting at adjacent points. Thus, the 'Lyapunov exponent' for a pseudo-chaotic system will stop growing near the average Hamming distance $N/2$ (Figure 21), where N is the highest possible Hamming distance.

Block ciphers, where the number of rounds is small (typically, $n = 16$), can be considered as chaotic systems. Long run pseudo-random generators have pseudo-chaotic properties (which resemble mathematical chaos).

Wolfram [50] describes a PRNG based on a one-dimensional cellular automata (CA). The system state is defined by a linear array of elements (cells)

$$\mathbf{b} = (b(1), b(2), \ldots, b(n)) \in \{0,1\}^n .$$

The iterated function $f : \{0,1\}^n \to \{0,1\}^n$ is given by

$$b(i) = b(i) - 1\, xor\, \big(b(i)\, or\, b(i+1)\big), \qquad \text{for all } 1 \le i \le n.$$

The array is considered circular i.e $b(n+1) = b(1)$, and the new element values are considered to be updated in parallel. The output is taken from one cell $b_k, k \in [0, n]$ only. This particular CA method requires three array-access operations plus two logical operations for each element of the array. While fairly simple to implement in hardware (since hardware may perform each cell computation in parallel), a large CA systems may involve more computation than desired for a software implementation using single CPU.

Gutowitz [18] suggests a complex scheme using cellular automata. A 512-bit plaintext block is encrypted into a slightly larger ciphertext (578 bits). A two part key (1088-bit) is used for a particular encryption scheme with so-called block-link structure. There are two rounds, each round consists of two subrounds and each subround includes substitution and permutation phases. The author expects that a sufficient number of rounds ensures a distribution that is statistically close to that of a random process and shows that the cipher is resistant to differential cryptanalysis.

Kolmogorov flows represent the most unstable class of chaotic systems, which can be particularly useful for mixing two dimensional data blocks. The iterated function T_ρ, also known as a generalized Baker map, can be considered as a geometrical transformation of a square image: the image is divided into vertical strips according to a partition set $\rho = \{n_1, n_2, \ldots, n_k\}$, stretched horizontally and compressed vertically as illustrated in Figure 22.

A discrete interpretation of the Kolmogorov transformation is based on modular calculus. The iterated function T_ρ is defined on a finite matrix:

$$T_\rho : \{0,1\}^N \to \{0,1\}^N$$

The set $\{n_1, n_2, \ldots, n_k\}$ is chosen in such a way that each integer n_i divides N and $n_1 + n_2 + \ldots + n_k = N$. The bit $b_{r,s} \in \{0,1\}^N$ is mapped to a new position

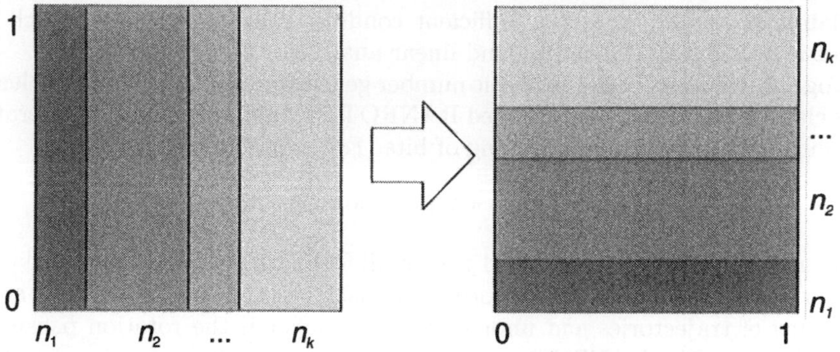

Figure 22. Stretch-and-compress transformation.

given by

$$T_\rho(r,s) = \left(\frac{N}{n_i}(r - N_i) + s \bmod \frac{N}{n_i}, \frac{n_i}{N}\left(s - s \bmod \frac{N}{n_i}\right) + N_i\right),$$

where $r, s \in \{1, 2, ..., N\}, N_i = n_1 + n_2 + ... + n_i, i \in \{1, 2, ..., k\}$ and $N_i \leq r < N_i + n_i$. Clearly, this discrete version of the transformation becomes a key-dependent cyclic permutation.

Scharinger [40] and Fridrich [13] and describe chaotic block ciphers based on the discrete Kolmogorov flows. The technique appears to be highly efficient since large matrices can be permuted in one round and relatively few iterations scramble the input object to unrecognizable form. However, this transformation should be used in combination with a substitution phase, since no permutation can change the statistical properties of the plaintext. With a simple modular addition Fridrich extends the transformation to three dimensions. The last 3D transformation is a good substitution cipher, which produces ciphertext with a uniform histogram in a few iterations. Diffusion phase is implemented with a Linear Feedback Shift Register (LFSR). The resulting encryption scheme is evaluated in terms of resistance to the general classes of attacks, and no cryptographic weakness is found. Cappelletti [6] design a hardware implementation (using FPGA) of the Scharinger's scheme.

Masuda *et al.* [32] suggest a cryptosystem based on a discretized one-dimension tent map

$$F(X) = \begin{cases} \left\lceil \frac{M}{A}X \right\rceil, & 1 \leq X \leq A \\ \left\lfloor \frac{M}{M-A}(M - X) + 1 \right\rfloor, & A < X \leq M \end{cases},$$

where $X, A = 1, 2, ..., M$ and $\lfloor\rfloor, \lceil\rceil$ round-down and round-up respectively. An initial condition is the plaintext; the final state is the ciphertext and a single parameter is the key. The chaotic map is applied sufficiently many times. The author discusses such properties as the correlation between invariant measures of plaintexts and ciphertexts. Sufficient conditions are given under which the cipher is resistant to differential and linear analysis.

Fog [12] describes pseudo-chaotic number generators with random cycle lengths. This class of chaotic systems (named RANROT) is similar to additive generators but with extra rotation or swapping of bits. For example,

$$x_n = \left((x_{n-j} + x_{n-k}) \bmod 2^b \right) \operatorname{rotr} r,$$

where $x_n \in \{0,1\}^b$, $r \in [0, b/2)$ and j, k are different integers chosen according to certain rules. The nonlinear transformation insures exponential divergence at the beginning of trajectories and unpredictable behavior if the rotation parameter r is unknown. The RANROT generators have passed all conventional tests of pseudo-randomness and appear to be faster than other generators of similar quality.

Classical iterative block ciphers can be considered as pseudo-chaotic ciphers because they prove out: (i) the sensitive dependence on initial conditions (in terms of plaintext and key) and (ii) the topological transitivity of states. For instance, the new Rijndael algorithm [38] accepted as AES[3] provides a 'full diffusion' after 2 rounds only. This block algorithm is based on N-round nonlinear substations and permutations. The S-box is a combination of multiplicative inverse and affine transformations (which is often used to generate chaos and fractals). The P-box represents cyclic row shifts and column mixing. Key expansion routines insure a key schedule generated from the cipher key. The resulting round key is XOR'ed with each block in each round.

6 Conclusions

This paper has discussed cryptography in the context of chaos theory. Clearly, all the conventional cryptographic systems and their components (iterative ciphers, PRNG, one-way functions etc.) represent pseudo-chaotic systems. In chaos theory and cryptography, we study a *nonlinear iterative transformation of the system state*, storing some useful information. Such properties as sensitivity to initial conditions, exponential divergence of trajectories, mixing can be applied in both areas. The difference that chaos theory focuses on the asymptotic behavior of unstable systems, while cryptography studies the effects of a finite number of iterations. Cryptographic pseudo-chaotic systems proves out a limited sensitivity to the initial conditions, i.e the exponential divergence of trajectories takes place during a finite number of iterations; after that one can notice a periodical behavior. Chaos is defined on the infinite state space; all computer implementations have a finite number of states. Unlike classical chaos, cryptographic systems have no recognizable attractor.

[3]The Advanced Encryption Standard is an encryption algorithm for securing sensitive but unclassified material by US Government agencies and, as a likely consequence, may eventually become the de facto encryption standard for commercial transactions in the private sector.

Chaotic continuous systems can, in principle, generate asymptotically random numbers. Floating-point or fixed-point arithmetic is the most obvious solution of approximating continuous chaos on a finite-state machine. However, this approach has some shortcomings which are unacceptable in digital cryptography. The output of such approximations of continuous chaos is not mathematically pseudo-random. The main problems are growing rounding-off errors and unpredictability short orbits.

Using one-step unpredictable chaotic systems with analytical solution, we can construct a generator and eliminate the precision problem in some degree: orbits become longer, but still, have different lengths.

Often, the iterated function or the analytical solution of a chaotic system rely on the same 'stretch-and-fold' mechanism suggested by Shannon. First, the seed is stretched over a large space (e.g, multiplying or raising in power), then folded M times into a smaller state space (using a periodical function such as mod, sin).

Yet, compared with binary systems, floating point approximations are inefficient and insecure. Consequently, a computer encryption algorithm should relay on binary chaos. Examples of such iterated functions are affine transformations [38], Feistel networks [41], linear and nonlinear feedback shift registers [25, 39], cellular automata [18], all of which are widely used in conventional cryptography.

Bibliography

1. Baptista, M. S., Cryptography with chaos. *Physics Letters A*, 240 (1–2), 1998, 50–54.

2. Beham, E., Cryptanalysis of the chaotic-map cryptosystem suggested at EUROCRYPT'91, 1991, `http://citeseer.nj.nec.com/175190.html`

3. Bianco, M. E., Reed, D., *An Encryption system based on chaos Theory*, US Patent No. 5048086, 1991.

4. Boffetta, G., Cencini, M., Falcioni, M., Vulpiani, A., *Predictability: a way to characterize complexity*, 2001,`http://www.unifr.ch/econophysics/`

5. Brown, R., Chua, L. O. Clarifying chaos: Examples and counterexamples, *Int. J. Bifurcation and Chaos*, **6**, 2, 219–249.

6. Cappelletti, L., *An FPGA implementation of a chaotic encryption algorithm*, Diploma Thesis, Universita Degli Studi di Padova, 2000, `http://www.lcappelletti.f2s.com/Didattica/thesis.pdf`

7. Carroll, J. M., Verhagen, J., Wong, P. T., Chaos in cryptography: the escape from the strange attractor, *Cryptologia*, **16**, 1, 1992, 52–72

8. Chu, Y. H., Chang, S., Dynamic cryptography based on synchronized chaotic systems, *Electronic Letters*, **35**, 12, 1999.

9. Chu, Y. H., Chang, S., Dynamic data encryption system based on synchronized chaotic systems, *Electronic Letters*, **35**, 4, 1999.

10. Dachselt, F., Kelber, K., Schwarz, W., *Chaotic coding and cryptanalysis*, 1997, `http://citeseer.nj.nec.com/355232.html`

11. Erber, T., Rynne, T., Darsow, W., Frank, M., The simulation of random processes on digital computers: unavoidable order, *J. of Computational Physics*, 49, 1983, 349–419.

12. Fog, A., *Chaotic random number generators*, 1999, `http://www.agner.org/random/theory/`

13. Fridrich, J., Symmetric ciphers based on two dimension chaotic map, *Int J. of Bifurcation and Chaos*, **8**, 6, 1998, 1259–1284.

14. Gallagher, J. B., Goldstein, J., *Sensitive dependence cryptography*, 1996, `http://www.navigo.com/sdc/`

15. Goldreich, O., *Introduction to Complexity Theory*, Lecture Note, Department of Computer Science and Applied Mathematics, Weizmann Institute of Science, Israel, 1999.

16. Golomb, S.W., *Shift register sequences*, Holden-Day, San Francisco, 1967

17. González, J. A., Pino, R., Chaotic and stochastic functions, *Physica*, 276A, 425–440.

18. Gutowitz, H., *Cryptography with Dynamical Systems*, ESPCI, Laboratoire d'Electronique, Paris, France, 1995, `http://www.santafe.edu/~hag/crypto/crypto.html`

19. Habutsu, T., Nishio, Y., Sasase, I., Mori, S., *A secret key cryptosystem by iterating chaotic map*, Lecture Notes in Computer Science, Advances Cryptology, Proceedings of EUROCRYPT'91, 1991, 127–140, `http://link.springer.de/link/service/series/0558/bibs/0547/05470127.htm`

20. Ho, M. K., *Chaotic encryption techniques*, Diploma Thesis, Department of Electronic Engineering, City University of Hong Kong, 2001, `http://personal.cityu.edu.hk/~50115849/ces/`

21. Hollasch, S., *IEEE Standard 754: floating point numbers*, 1998, `http://research.microsoft.com/~hollasch/cgindex/coding/ieeefloat.html`

22. Hosack, J., The use of Chebysev mixing to generate pseudo-random numbers, *Journal of Computational Physics*, 67, 1986, 482–486.

23. Jackson, E. A., *Perspectives in nonlinear dynamics* , vol. 2, Cambridge Univ. Press, Cambridge, 1991, 33.

24. Kennedy, M. P., Rovatti, R., Setti, G., *Chaotic electronics in telecommunications*, CRC Press, 2001, `ISBN 0-8493-2348-7`

25. Knuth, D., *The art of computer programming - seminumerical algorithms*, vol. 2, 2nd ed. Addison-Wesley: Reading, Massachusetts, 1981.

26. Kocarev, L. J., Halle, K. S., Eckert, K., Chua, L. O., Experimental demonstration of secure communications via chaotic synchronization, *Int. J. Bifurcation and Chaos*, 2(3), 709–713.

27. Kocarev, L., Chaos and cryptography, 2001, `http://rfic.ucsd.edu/chaos/ws2001/kocarev.pdf`

28. Z.Kotulski, J.Szczepański, Discrete chaotic cryptography. New method for secure communication, *Proc. NEEDS97*, 1997, `http://www.ippt.gov.pl/~zkotulsk/kreta.pdf`

29. Kotulski, Z., Zczepanski, J., On the application of discrete chaotic dynamical systems to cryptography. DCC method, Biuletyn Wat Rok, **XLVIII**, 10 (566), 1999, 111–123, `http://www.ippt.gov.pl/~zkotulsk/wat.pdf`

30. Kotulski, Z., Szczepański, J., Grski, K., Grska, A., Paszkiewicz, A., On constructive approach to chaotic pseudorandom number generators, *Proceedings of the Regional Conference on Military Communication and Information Systems*, CIS Solutions for an Enlarged NATO, RCMIS 2000, October, Zegrze, **1**, 191–203, `http://www.ippt.gov.pl/~zkotulsk/CPRBG.pdf`

31. Lai, D., Chen, G., Hasler, M., Distribution of the Lyapunov exponent of the chaotic skew tent map, *Int. J. of Bifurcation and Chaos*, **9**, 10, 1999, 2059–2067.

32. Masuda, N., Aihara K. [2000], *Finite state chaotic encryption system*, 2000, `http://www.aihara.co.jp/rdteam/fs-ces/`

33. Matthews, R., On the derivation of a chaotic encryption algorithm, *Cryptologia*, 13, 1989, 29–42.

34. Menezes, A. J., Oorschot, P. C. van, Vanstone, S. A., *Handbook of applied cryptology*, CRC Press, 1996, `http://www.cacr.math.uwaterloo.ca/hac/`

35. , *A statistical test suite for the validation of random number generators and pseudo random number generators for cryptographic applications*,2001, `http://csrc.nist.gov/rng/rng2.html`.

36. Paar, N., Robust encryption of data by using nonlinear systems, 1999, `http://www.physik.tu-muenchen.de/~npaar/encript.html`

37. Protopopescu, V. A., Santoro, R. T., Tolliver, J. S.,*Fast and secure encryption-decryption method based on chaotic dynamics*, US Patent No. 5479513, 1995.

38. Rijmen, V., Daemen, J., *Rijndael algorithm specification*, 1999, `http://www.esat.kuleuven.ac.be/~rijmen/rijndael/`

39. Ritter, T., The efficient generation of cryptographic confusion sequences, *Cryptologia*, 15, 1991, 81–139.

40. Scharinger, J., *Secure and fast encryption using chaotic Kolmogorov flows*, Johannes Kepler University, Department of System Theory, 1998, `http://www.cast.uni-linz.ac.at/Department/Publications/Pubs1998/Scharinger98f.htm`

41. Schneier, B., *Applied cryptography, second edition*, John Wiley & Sons, Inc, 1996, ISBN 0-471-12845-7

42. Shannon, C. E., A mathematical theory of communication, *Bell System Technical Journal*, **27**, 4, 1948, 379–423, 623–526.

43. Shannon, C. E., Communication theory of secrecy systems, *Bell System Technical Journal*, **28**, 4, 1949, 656–715.

44. Sobhy, M. I., Schehata, A. E. D., Secure e-mail and databases using chaotic encryption, *Electronic Letters*, **36**, 10, 2000.

45. Svensson, M., Malmquist, J. E., *A simple secure communications system utilizing chaotic functions to control the encryption and decryption of messages*, Project report for the course 'Chaos in science and technology', Lund Institute of Technology, Dept. of physics, Subdept. of mathematical physics, 1996, `http://www.efd.lth.se/~d92ms/chaoscrypt.html`

46. J.Szczepański, Z.Kotulski, Pseudorandom number generators based on chaotic dynamical systems, *Open Systems & Information Dynamics*, **8(2)**, 2001, 137–146, `http://www.ippt.gov.pl/~zkotulsk/open.pdf`

47. Waelbroeck, H., Zertuche F., *Discrete chaos*, 1998, `http://papaya.nuclecu.unam.mx/~nncp/chaos98.ps`

48. Wheeler, D.D., Problems with chaotic cryptosystems, *Cryptologia*, 12, 1989, 243–250.

49. Wheeler, D. D., Supercomputer investigations of a chaotic encryption algorithm, *Cryptologia*, 15, 1991, 140–150.

50. Wolfram, S., Random sequence generation by cellular automata, *Advances in Applied Mathematics*, 7, 1986.

51. Wong, W. K., *Chaotic Encryption Technique*, City University of Hong Kong, Department of Electronic Engineering, Hong Kong, 1999, `http://kitson.netfirms.com/chaos/`

52. Wu, C. J., Lee, Y. C., Observer-based method for secure communication of chaotic systems, *Electronic Letters*, **36**, 22, 1999.

The Making of "Fractal Geometry in Digital Imaging"

Martin J. Turner and Jonathan M. Blackledge
ISS, SERC, Hawthorn Building, De Montfort University, Leicester LE1 9BH

1 Introduction

The Institute of Simulation Sciences is concerned with research into the simulation of engineering systems and data processing. The institute has a range of activities which include the *centre for digital signal and image processing.*

http://www.serc.dmu.ac.uk/ISS

This centre is concerned with algorithms for digital signal processing with applications in digital communications and networks, mobile radio communications and encryption. Research into image compression, image synthesis, image processing for remote sensing, non-destructive evaluation and computer vision systems is sponsored by industries that includes British Telecom, Marconi and the Defence Evaluation and Research Agency (DERA). The centre has a particular strength in the application of fractal geometry to image synthesis and computer graphics in general. Novel mathematical approaches are being considered for modelling non-stationary self-affine stochastic fields with applications to digital printing and financial analysis. This has led to the development of two new commercial products called Microbar (TM) and Micromark (TM) respectively, which are being marketed by Microbar Security Limited.

During the IMA conference there were presented some of the fractal images taken from the published book **'Fractal Geometry in Digital Imaging'** [ISBN: 0127039708 Academic Press, 1998] in a series of posters, as well as a VHS video showing some of the animations. Within the following text are descriptions of the making of some of the less obvious images. Other books published recently by members of the institution are:

Image Processing III Jonathan M. Blackledge and Martin J. Turner editors from the "IMA Third Image Processing: Mathematical Methods, Algorithms and Applications": Based on the Proceedings of the Third IMA Conference on Image Processing. 2001 Horwood Publishing; ISBN: 1898563721

Image Processing II Jonathan M. Blackledge and Martin J. Turner editors from the "IMA Second Image Processing: Mathematical Methods, Algorithms and Applications": Based on the Proceedings of the Second IMA Conference on Image Processing. 2000 Horwood Publishing; ISBN: 1898563616

Analytic Methods for Partial Differential Equations (Springer Undergraduate Mathematics Series) by G. Evans, J.M. Blackledge, P. Yardley Two Volumes. Hardcover (November 1999) Springer Verlag; ISBN: 3540761241

Delta Functions - Introduction to Generalised Functions by Roy F. Hoskins, Horwood Series in Mathematics & Applications. Ellis Horwood Publishing Limited; ISBN: 1898563470

2 Images of the Book

The vast majority of the layout and image creation of the book *Fractal Geometry in Digital Imaging* was carried out on a small SGI Indy workstation; MIPS R4400 (150 MHz) 128 MB (Indy 24 bit), 3GB Hard discs, Floptical removable storage and Vino video card. Within the covers of the final book there are over 90,000 words and over 520 incorporated images and diagrams.

To create all these images, numerous pieces of software were used, and within this short paper it is hoped to explain some of the tools as well as present a few asides on the creation of certain specific figures. The text and layout were constructed within the LaTeX typographical language, using associated packages for the bibliography and index. To quote Donald Knuth from has famous volume, *Fundamental Algorithms - The Art of Computer Programming*, regarding the index he commented that "any inaccuracies ... may be explained by the fact that it has been prepared with the help of a computer". This can now be extended to almost all parts of the book; including keeping track of figures, page numbers, glossary entries and bibliography references as well as the index. The final printing at Academic Press required the creation of individual files for different parts on a CD Recordable disc, before direct printing.

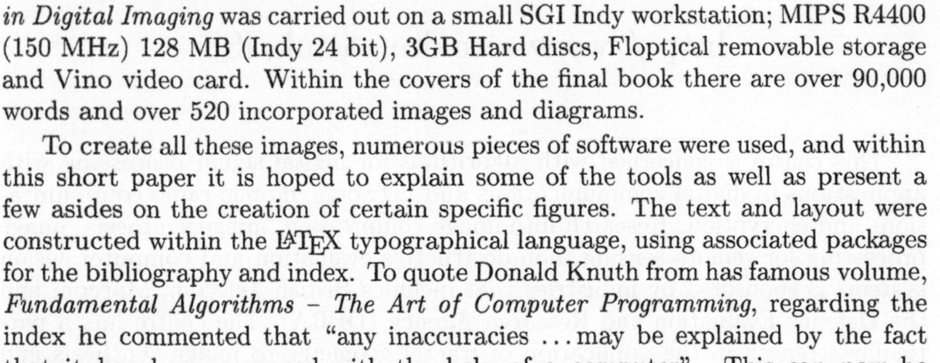

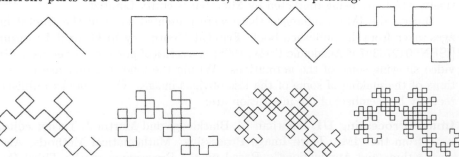

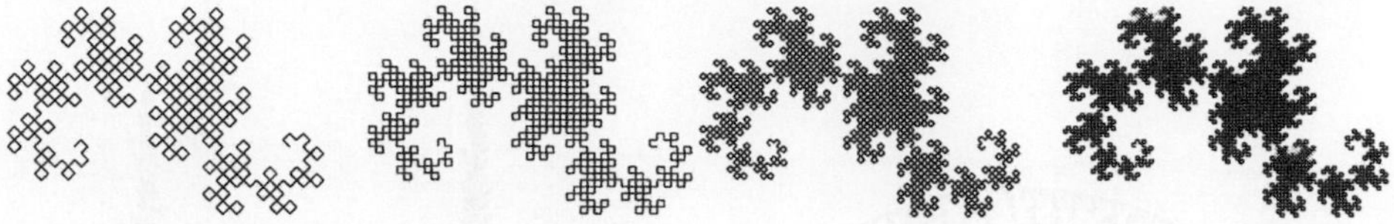

The images were translated into PostScript format before being incorporated into the text. Virtually all the stick drawings were constructed within a UNIX product called `idraw`. It is not a product of choice, being cumbersome to install and missing lots of features that are found on more universal drawing packages, for example `Illustrator` from Adobe, but it is freely available on a UNIX platform and does produce fairly simple PostScript output. PostScript is a graphical programming language developed by Adobe mainly used by printers and achieves (almost) resolution independent output. This means that the self-similar fractal images, for example the Herter-Heighway dragon curve, shown above, can be constructed by programming the rules directly. Thus the laser printer will take a description of a line and repeatedly duplicate, scale and align it to as many levels as required, and will then print the final image with as high a resolution as the current printer can handle. This can lead to very slow printing time; for example, with high resolution printing eleven generations of the curve could be created but this took over three minutes to print on an HP-LaserJet 4. This does mean that often what is seen on the pages has been constructed by a processor using the rules that are described within the text. In the following, we discuss and show some of the images and briefly address the way in which they have been created.

Three DLA (diffusion limited aggregation) images are shown below. The image on the left is a race where particles are continually inserted into the centre of the circle and progressively wander around in a random walk. When a particle reaches the edge of the circle or another particle that is already stationary it sticks.

There is a 'competition' to see which part of the circles perimeter will reach the centre first, whereupon the process terminates. As pixels are in this simulation represented as squares there is a fractional North-South-East-West bias in this competition. The angle location of intersection would create an unusual, and extremely slow, random number generator as it is the sum of many random numbers.

In the two image on the right particles are released from the top with a predefined force of gravity giving them a strong bias downwards. The spread of the particles varies as the force of gravity is changed. In normal Brownian motion DLA, the fractal dimension is defined as 1.5, but by modifying the motion characteristics and the world geometry a whole range of fractal dimension attractors can be created and thus a whole range of synthetic textures. Simulations of flat coral to fragile ferns can be created in this way.

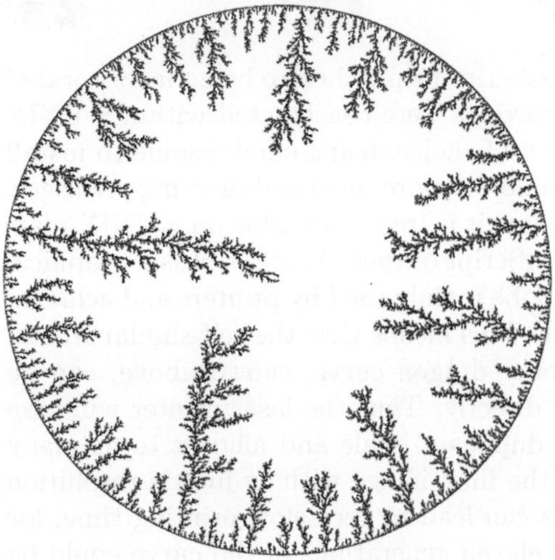

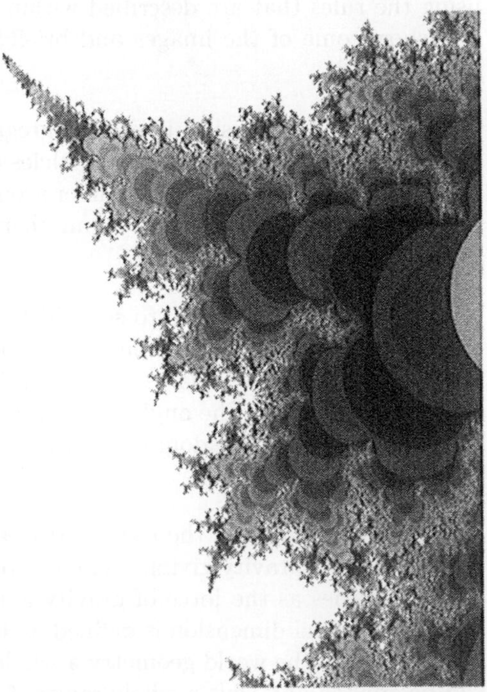

A classic zoomed in view of the famous Mandelbrot set. Each point on the image represents a complex number $c = x + yi$, that defines the iteration; $z_{j+1} = z_j^2 + c$. Starting with $z_0 = 0$ we colour the location c dependent on the value of n, such that the first value of z_n is less than a given threshold value. This image uses a standard threshold value of 2.

The strict relationship between fractal types offers a method of analysing certain features of a fractal. A Julia set can be created at each point in the Mandelbrot set and within the two images below various points have been constructed. The transitions from one Julia set to another within the Mandelbrot set is continuous and features for example; scaling, connectivity and number of internal spirals can be easily seen. This can be observed from the various selection of points within the images below.

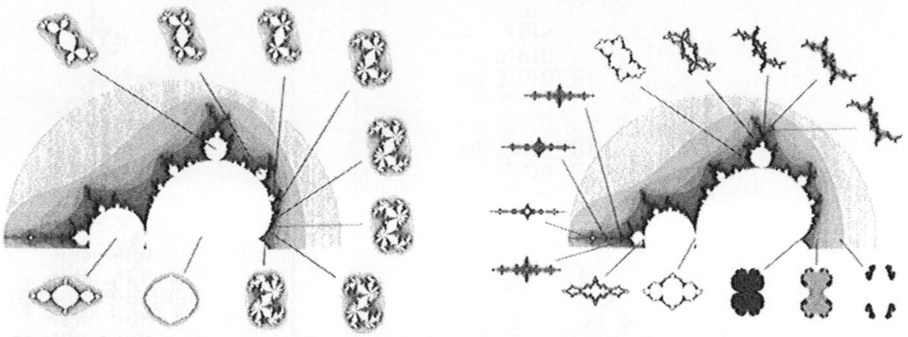

Names for these locations have been invented, possibly due to over-imagination, including, rabbit, dragon, dentrite and chrysanthemum. Sequences have also been discovered, with occasional uses. The mini-images in the images above are Julia curves at different positions within the Mandelbrot set. The series on the left indicate locations within major holes or gaps of the Mandelbrot set with resulting sub-images each branching at an increasing rate as the holes decrease in size. The top-left image branches with a factor of two and is sometimes referred to as the *rabbit.* All these sequences can, and have many times, been used to create smooth animations. Simple points traveling horizontally through the axis of the Mandelbrot image and a second series of points traveling down a *channel* are shown in the image on the right.

A series of recursive patterns created by removing one or more of the edges of the Peano generator using an L-system (Lindermayer systems). The following diagrams show the generators and the resulting image created when one, two or three lines of the initiator are removed, creating various doilly type patterns. L-systems are a great way of creating complex objects from simple descriptors and are often found within 3D modelling packages; although easy to program directly.

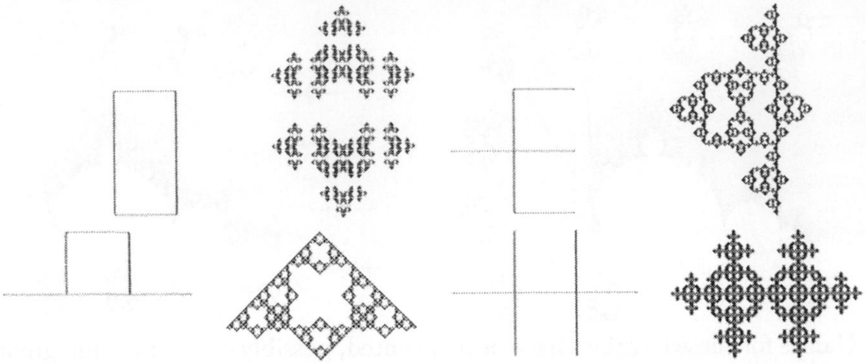

The following are not true fractal but an interesting example of creating something very complicated out of a very simple function. Each point in the image (x, y) represents an integer created from the function, $y \times x^2$. The black and white images are created by examining the nth bit of the binary number to see if it is zero or one. When the value is one, a black pixel is drawn. The first image represents the 11th bit on an image array, 700×1059, and the second image represents the 19th bit on an image array, 1500×2300.

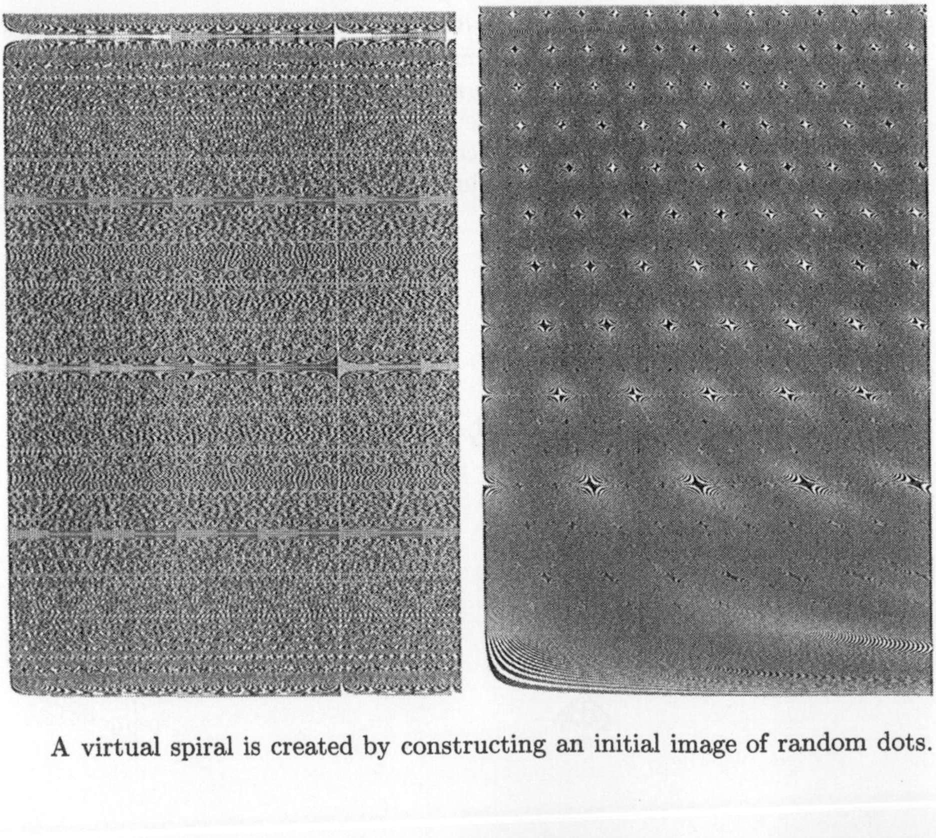

A virtual spiral is created by constructing an initial image of random dots.

This image is then copied, rotated by 3 degrees and finally scaled from the middle by 10 percent before being overlayed on top of the initial image. The image created here shows the overlay of three copies giving the visual appearance of a spiral. The further away the points are from the centre the harder it is for the human visual system to see the spiral. The other two images are simple superposition of a single rescaled or rotated random dot image.

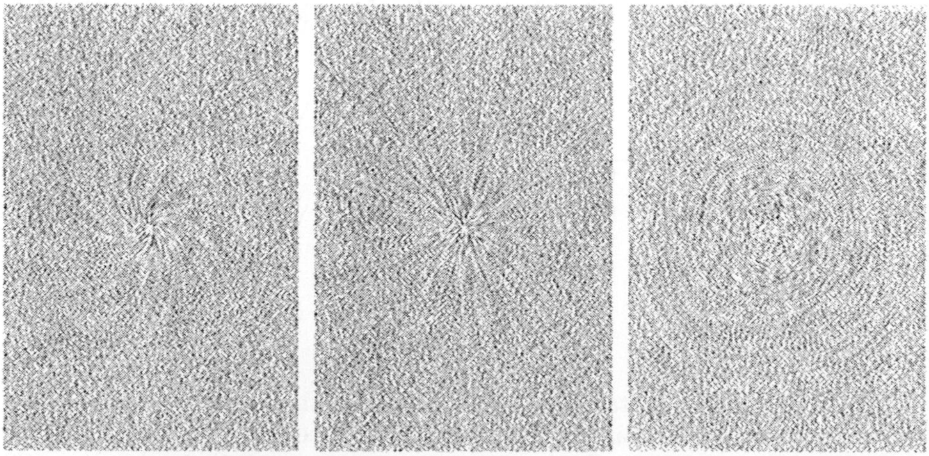

Also there are many image that creates chaotic behaviour out of sampling. Each location has a calculated value;

$$z(x, y) = (\text{Trunc}\,\{\alpha(\sin x + \sin y)\}) \bmod 3$$

When $z(x, y) = 0$ then a black dot is drawn, x running vertically and y horizontally. The value of α varies from zero at the left of the page to 120 at the right. This formula comes from Clifford Pickover's book 'Computers Patterns Chaos and Beauty'. Resolution in floating point notation plays an important role in the resulting pattern.

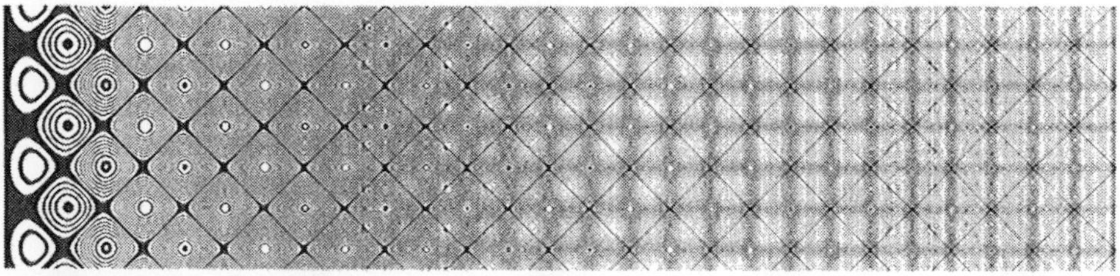

The above images can be considered in some way as manipulations of an Iterated Function System. Presented below is a selection of frames that linearly interpolate between two IFS transforms; from the

Sierpiński triangle;

a_i	b_i	c_i	d_i	e_i	f_i
0.5	0	0	0.5	0	0
0.5	0	0	0.5	100	0
0.5	0	0	0.5	50	50

to Barnsley's fern.

a_i	b_i	c_i	d_i	e_i	f_i
0.85	0.04	−0.04	0.85	0	1.6
0.2	−0.26	0.23	0.22	0	1.6
−0.15	0.28	0.26	0.24	0	0.44
0	0	0	0.16	0	0

As Barnsley's fern contains one extra transform, the fourth transform has to be created from nothing as the animation commences. Again, this shows the smooth continuous nature of the attractor image given a slowly changing IFS definitions. Lots of fractal signal definitions have been used for image animation sequences, by slowly modifying one or more of the parameters. Shown below is a cloud vanishing, created using Maya from Alias|Wavefront, with the raycasting tools on a fractal transparency disk overlaying a perfectly formed oval cloud.

and a stylised explosion,

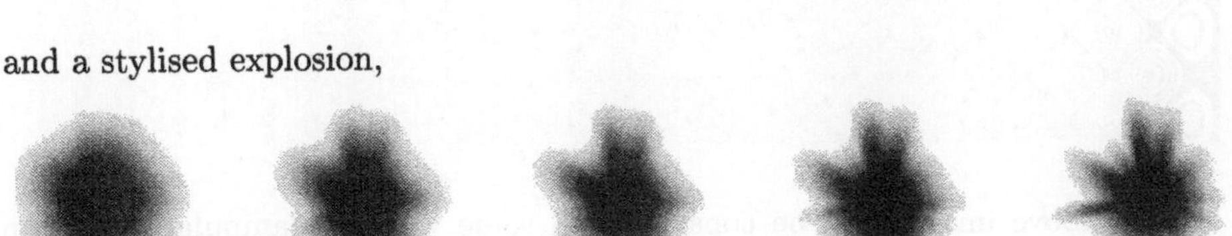

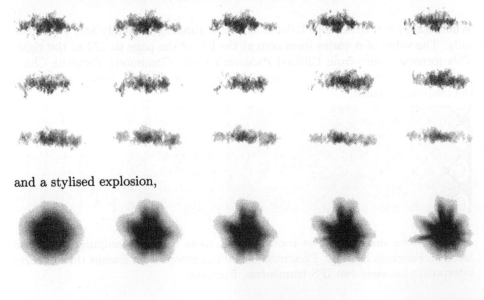

To contrast this the following three IFS transforms (attractors shown below left) are very similar to each other, but create very different images. The bottom one has been duplicated, and they have all been rotated to fit as a collage. The top image is defined by the following transforms;

i	a_i	b_i	c_i	d_i	e_i	f_i
1	−0.8	−0.5	−0.4	0.9	200	200
2	−0.2	−0.1	−0.1	−0.2	0	0

the bottom image is defined as;

i	a_i	b_i	c_i	d_i	e_i	f_i
1	0.8	0.5	0.4	−0.9	200	200
2	−0.2	−0.1	−0.1	−0.2	0	0

and the spiral in the middle is defined as:

i	a_i	b_i	c_i	d_i	e_i	f_i
1	−0.8	0.5	−0.4	−0.9	200	200
2	−0.1	−0.1	−0.1	−0.1	0	0

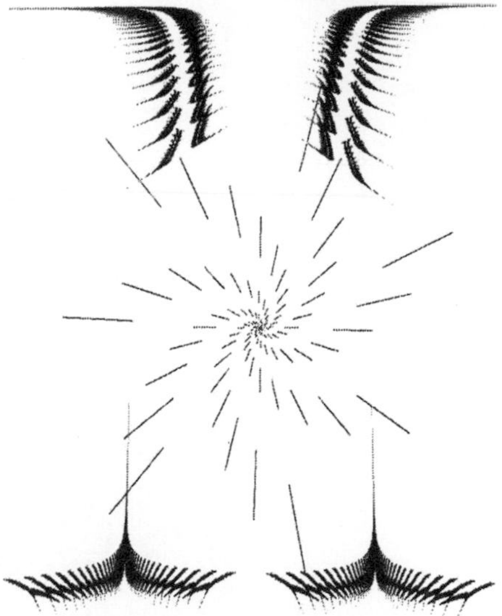

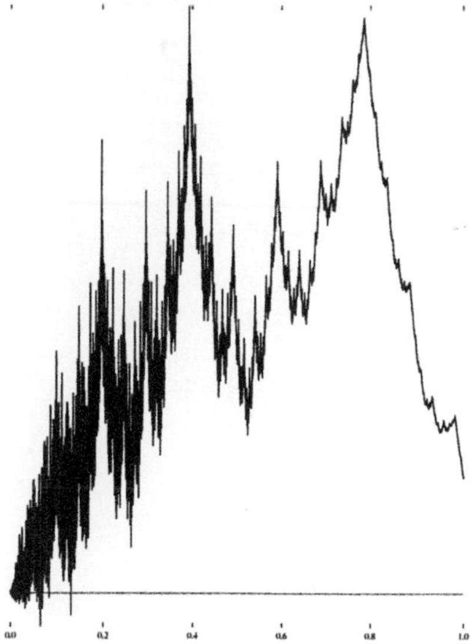

Finally, shown above right, is a fractal signal, based on the Weierstraß function, that has a continually varying fractal dimension. Signals like this are very useful for analysing and testing fractal dimension extracting algorithms. This one has the wonderful property of having a local dimension at position x of $2-x$.

$$f(x) = x^{3/2} \sum_{n=0}^{\infty} 2^{-nx} \cos\left(2^n x\right)$$

Printed and bound by CPI Group (UK) Ltd, Croydon, CR0 4YY
03/10/2024
01040435-0008